GREEN ROOFS, FACADES, AND VEGETATIVE SYSTEMS

Safety Aspects in the Standards

GREEN ROOFS, FACADES, AND VEGETATIVE SYSTEMS

Safety Aspects in the Standards

ELENA GIACOMELLO

Università Iuav di Venezia
Iuav University of Venice, Italy

ELSEVIER

Butterworth-Heinemann
An imprint of Elsevier

Butterworth-Heinemann is an imprint of Elsevier
The Boulevard, Langford Lane, Kidlington, Oxford OX5 1GB, United Kingdom
50 Hampshire Street, 5th Floor, Cambridge, MA 02139, United States

Library of Congress Cataloging-in-Publication Data
A catalog record for this book is available from the Library of Congress

British Library Cataloguing-in-Publication Data
A catalogue record for this book is available from the British Library

ISBN: 978-0-12-817694-8

For information on all Butterworth-Heinemann publications visit our website at https://www.elsevier.com/books-and-journals

Publisher: Matthew Deans
Acquisitions Editor: Carrie Bolger
Editorial Project Manager: Peter Adamson
Production Project Manager: Kamesh Ramajogi
Cover Designer: Matthew Limbert

Typeset by TNQ Technologies

Contents

CHAPTER 1

Introduction

The research *Green Roofs, Façades, and Vegetative Systems: Safety Aspects in the Standards* is an extensive analysis of 25 documents related to the design of green roofs and vegetative façades/walls, with a view to identify the quality and effectiveness of the requirements and instructions for vegetative technology and systems.

A "vegetative system" is a technological system onto which vegetation is intentionally grown.

Among several layers, it includes vegetation, growing media and water (irrigation or storm water) represent the "alive part" of the technology, characterized by a changing and adaptable behavior related to climate and microclimate and their elements (and factors), exposure to sun/shadow, and other conditions, for example, possible plant disease.

The vegetative systems for roofs and façades are named in different ways, which most of the time may be considered as synonymous: green roofs/walls, roofs gardens, living roofs/walls, etc. In this book, we will define all of them as "vegetative systems."

In recent years, vegetative systems have taken advantage of the newest technical advances in the field of agronomy and have applied them to the design of architecture: many vegetative systems for roofs have been improved, including modular green roofs, biodiverse green roofs, vegetable roof gardens, brown roofs, extremely light and weight green roofs, etc., and, in the last 20 years, many different vegetative façades systems—modular panels, cladding plant-container façades, revisited climbing plants façades, hydroponic walls, etc., prevegetated and not—have been continuously developed, tested, and applied.

In the meantime, the application of vegetative systems to buildings and tall buildings has also become more popular, due to the appreciation from building occupants and marketability for developers. Moreover, the positive reception from general public and city governments, when these systems are applied to buildings, has encouraged designers to incorporate the most advanced vegetative elements in and around new buildings: on the envelope and balconies, in interiors, around bearing structures, inside and outside tall buildings surfaces, even along infrastructure, and so on, often

Green Roofs, Facades, and Vegetative Systems
ISBN 978-0-12-817694-8
https://doi.org/10.1016/B978-0-12-817694-8.00001-7

with spectacular effects when applied in dense city contests and mimetic results when applied in natural and rural environment.

Considering the great success of vegetative systems and the consequent wide application in different contests, also unconventional, recently, developers and insurance companies have begun to ask serious questions about the safety of vegetative systems: are the new vegetative systems applied to buildings safe?

As vegetative roofs and façades are hybrid technologies, consisting of organic materials and traditional construction materials, can vegetative roofs and façades pose a fire hazard?

Further questions raised are as follows: are vegetative systems susceptible to wind forces? Which documents provide information on designing vegetative building envelopes to resist to wind?

What are the reference standards for vegetative systems applied to buildings and tall buildings?

Since the beginning of 80s, the most important reference document (a guidelines, not a standard!), widely used at global scale, has been the German FLL-Green Roofs Guidelines,[1] which provides total requirements for designing, applying, and maintaining green roofs to buildings.

Then, with the expansion of the market, new standards, regulations, and guidelines appeared in several States: this gave an important impulse to a "safer market" of vegetative systems expanding and somehow formalizing the state-of-the-art vegetative systems worldwide.

To answer the important questions regarding the safety of vegetative systems, in exhaustive manner, an analysis of the entire standard corpus on vegetative systems, published at global scale, has been conducted.

Of the 25 documents examined, 7 are standards developed exclusively for vegetative systems. The remaining 18 documents are handbooks, building codes, city codes, etc., which were included in this research to understand if standards and other technical documents provide enough information to ensure that safety is achieved in the design, construction, and maintenance of green roofs and green façades/walls, or if they are outdated or insufficient, in particular for what concerns safety aspects. It is now the time to ask if the standards are still applicable to the newest vegetative systems or if they need to be expanded.

Endnote

1. See §Bibliography.

CHAPTER 2

Research steps

Objectives

The main research objective was to analyze the international standards for vegetative systems and address the following concerns for four main topics inherent safety and risks of vegetative systems:

(1) Fire risk/fire safety: Considering that green roofs and façades/walls partly consist of organic components (plants and substrates) and that organic materials are flammable, it is fundamental to define their fire safety. Are vegetative systems a fire hazard? Which documents provide information on designing these systems for fire resistance and fire safety? Are the guidelines, requirements, and standards outlined in these documents enough to ensure adequate fire safety?

(2) Structural Safety and wind Design: Intensive vegetative systems, which include shrubs and trees, are successfully applied to roofs and façades of "regular buildings" and also tall buildings. Are vegetative systems susceptible to wind forces? Are retaining systems for plants different according to plant dimensions and plant species?

(3) Irrigation: Most (not all) vegetative systems are irrigated to guarantee hydration to plants in specific conditions. Can irrigation be used as a method to combat fire and to guarantee fire resistance to roofs and façades?

(4) Maintenance/cost estimation: The regular maintenance is the most important action to ensure life, health, and safety to green roofs and vertical green. Do the standards make the topic enough explicit? Is the information to conduct a correct maintenance complete? Do the guidelines, requirements, and standards provide enough information to ensure that the developer, building manager, designer, etc., are aware of all costs related to the maintenance of each green system? What effect has maintenance on the overall safety of the vegetative systems?

Green Roofs, Facades, and Vegetative Systems
ISBN 978-0-12-817694-8
https://doi.org/10.1016/B978-0-12-817694-8.00002-9

All these questions require technical and precise answers. This research aims to determine if and which documents provide information to ensure all safety measures are met for the four topics examined (i.e., fire risk, wind design, irrigation, and maintenance). This analysis will identify the standards, guidelines, research reports, etc., that provide answers to these questions and the topics that require further clarifications or research.

Methodology

The research was conducted in an analytical process, with a view to identify how and to what extent the four topics are addressed in each document. This analytical process was conducted with the following research actions:

Identification and selection of documents

The selection of documents was conducted through suggestions from the peer review team of around 20 experts in the green building/sustainability, fire safety, landscaping, tall building, and wind design industry and through research conducted by the research team.

The analyzed documents are 25: 7 standards (exclusively developed for green roofs) and also "official documents" most of which provide extensive information to ensure that safety is achieved in the design and construction of vegetative roofs and vegetative façades/walls.

In addition to standards—which are the most important documents—the mentioned "official documents" include the following types: building codes, safety codes and city codes and guidelines, handbooks, manuals, research reports undersigned from national/international recognized institutions, authorities, and associations.

All of the documents suggested by the peer reviewers were included and analyzed in this research. Although some of the topics discussed in the suggested documents were outside of the scope of this research, the documents did contribute to enrich the overall knowledge of the topics and identified areas where new considerations should be taken, specifically the integration of requirements for vegetative systems into standards and building codes.

Document analysis

Each document has been analyzed and summarized with the following contents:

(1) A brief description of the document and identification of the system and/or topics that are addressed;

(2) The structure of the document (number and name of sections, chapters, appendixes, etc.);

(3) Peculiarities, strengths, and limitations of the documents, if any;

(4) Identification of the sections within each document that address each of the four topics (i.e., fire risk/safety, structural/wind design, irrigation, maintenance).

Extraction of the documents contents and creation of a summary table

Following the identification of the sections in each document where the fire risk/safety, structural/wind design, irrigation, and maintenance are discussed, an extraction of the information was performed in two phases:

(1) Direct extraction of the contents of each document

The direct extraction of the contents of each document was performed to provide, with the exact wording, technical details of each document. This task was performed to provide an easier method to directly reference the exact contents of each of the document, where the four topics are discussed.

(2) Creation of a summary table

The creation of a summary table was performed to share the research findings of the research report in a clear and concise manner. This is done to identify the standards, guidelines, codes, etc., that are available and to identify which of these documents provide extensive (or limited) information for each topic.

The table is organized to provide information about each document in a readable and simple manner. The information provided for each document includes the following:

- The type of document (i.e., standard, code, research report, etc.);
- Document information (i.e., title, authors, number of pages, publication year);
- Whether the document focuses on vegetative façades, roofs, or both;

- Which specific chapters, sections, tables, etc., provide information on each of the four topics, if applicable;
- The brief description of the contents of the chapters, sections, tables, etc., (if any) where the four topics are discussed.

The table is organized as follows:

n°	Document	Roof/ Façade	Fire risk/Fire safety	Structural safety/ Wind design	Irrigation	Maintenance/ Cost estimation
Type of documents (i.e., standard, code, research report, paper, etc.)						
n°	Author, Title, number of pages, year	Roof or façade	Identify the sections in each document where the respective topics are discussed and provide a summary of the content of the sections.			

The aim of this is to provide the reader with a clear method to identify where and to what extent each topic is discussed, without compromising the intellectual property of each document.

CHAPTER 3

Preliminary considerations

Of all analyzed documents, the following main considerations can be made:

(1) Through the extensive analysis of each document, it became clear that the mention of vegetative façades/walls in existing documents devoted to vegetative technologies is rare; vegetative façades/walls are only extensively discussed in two of the documents analyzed. The primary scope of this research was to analyze standards, and at this time no standards have been developed for vegetative façades/walls. This could be due to a variety of reasons: compared to green roofs, vegetative façades/walls are a more rare and recent design feature; there is a broad range of different types of vertical vegetative systems (i.e., climbing plants, hydroponic walls, vertical green wall panels, vegetation on plant containers along the façades, etc.), making it much more difficult to create a standard that provides information for all aspects of each system and for different climate; new types of systems are still being studied, tested, designed, and produced. For this reason, this research focuses on the concerns of green roofs, except for where is explicitly stated otherwise;

(2) As stated, this research focuses primarily on the analysis of standards, as they identify the state-of-the-art vegetative systems that have been deeply researched and developed. Additional documents (i.e., codes, research reports, etc.) have been analyzed to identify aspects of the topics that may not be considered or developed enough in the standards;

(3) Like vegetative façades/walls, the application of vegetative systems on tall buildings (where safety requirements are normally more and more restrictive) is another topic that is not thoroughly discussed in the analyzed documents. References to buildings over 45.7 m (150 ft) are only made in two of the documents: ANSI/SPRI VF1 and FM Global Property Loss Prevention Data Sheets;

(4) The type, the scope, the language, and the structure of the documents significantly differ. This analysis tries to make comparable all the texts, without loosing information and meaning in the synthesis of

Green Roofs, Facades, and Vegetative Systems
ISBN 978-0-12-817694-8
https://doi.org/10.1016/B978-0-12-817694-8.00003-0

information drawn from each document. The following frames the main characteristics of each document, separated by type (i.e., standard, guideline, handbook, etc.).

Standards

7 standards were examined.[1]

The first two standards examine all the requirements of the system layering of a green roof and the related topics for designing, constructing, and maintaining a green roof. As many European standards, both of them are performace oriented:

Standard (1)

UNI-Ente Italiano di Normazione, 2015, UNI 11235:2015 Istruzioni per la progettazione, l'esecuzione, il controllo e la manutenzione di coperture a verde (tr. Criteria for design, execution, testing and maintenance of roof gardens).

UNI 11235:2015 is a standard regarding the design, construction, and maintenance of green roofs. It may be defined as a "full standard," as it provides requirements and performance for each layer of the entire system. The first version was released in 2008 and the second version was revised in 2015. This standard is partly inspired by the FLL Green Roofs Guidelines[2] but was developed to address requirements and performance for green roofs installed in warmer climates, like Mediterranean ones. One major limitation of UNI 11235 is that, at the moment, it is released only in Italian.

[1] Two standards inherent green roofs, from Austria and Switzerland, are not analyzed in this publication: (1) ÖNORM L 1131:2010 06 01 Gartengestaltung und Landschaftsbau—Begrünung von Dächern und Decken auf Bauwerken—Anforderungen an Planung, Ausführung und Erhaltung—Garden design and landscaping—Greening of roofs and ceilings on structures - Requirements for planning, execution and conservation. 59 pages; (2) SN-Schweizer Norm 564 312, SIA 312:2013 Begrünung von Dächern—Roofs Greening. 24 pages.

[2] The FLL Green Roofs Guidelines is the benchmark document worldwide for the design, construction, and maintenance of green roofs. The industrialization of vegetative systems for roofs started in Germany, where the state of the art was based around 40 year ago with the publication of these guidelines, translated in English and revised in the time. Up to the present day, the FLL Green Roofs Guidelines represents the reference document for green roofs at global scale.

Standard (2)

MCCAA-Malta Competition and Consumers Affairs Authority, Standards and Metrology Institute, SM 3700:2017, Green Roofs—Criteria for the planning, construction, control and maintenance of Green Roofs.

The Maltese Standard SM 3700:2017 is an extended document which provides all requirements for designing, constructing and maintaining green roofs. As the Italian Standard it is devoted to Mediterranean climates.

UNI 11235:2015, SM 3700:2017 may be considered as analogue documents, as they cover all aspects necessary to design a green roof. Although they reference other national standards for specific requirements of some of the green system materials, components, structures, etc., they represent the most complete standards for green roofs design, construction, and maintenance.

Standard (3)

ASTM International—American Society for Testing and Materials International, 2014, ASTM E2777 Standard Guide for Vegetative (Green) Roof Systems.

The ASTM E2777-14 is an introductory standard for the design of green roofs. It is a useful document, providing extensive explanations about green roof concepts and useful other information (i.e., terminology, reference documents, general principles, and explanations of main technical requirements), but no prescriptive dimensional information is provided.

Standard (4)

ANSI/SPRI—American National Standard Institution/Single Ply Roofing Industry, 2017, ANSI/SPRI VF-1 External Fire Design Standard for Vegetative Roofs.

This standard provides information to design fire-resistive green roofs. It refers to ASTM E108 "Standard Test Methods for Fire Tests of Roof Covering" for class A materials.

The standard specifies general requirements to ensure the fire safety of green roofs (i.e., system requirements, walls and fire break roof areas, border zones, etc.).

Standard (5)

ANSI/SPRI—American National Standard Institution/Single Ply Roofing Industry, 2016, ANSI/SPRI RP-14, Wind Design Standard for Vegetative Roofing System.

The ANSI/SPRI RP-14 provides a method of designing wind uplift resistance of vegetative roofing systems utilizing adhered roofing membranes. It provides a procedure for defining the wind requirements of green roof systems and the type of ballast needed, based off of a set of measurable conditions.

Standard (6)

CSA Group—Canadian Standards Association, 2015, A123.24-15 Standard test method for wind resistance of modular vegetative roof assembly.

The A123.24-15 is a standard that provides methods for testing the wind flow resistance and wind uplift resistance of modular green roofs. The standard defines modular green roofs as a modular vegetated system (a system consisting of pregrown or precultivated vegetation: modules, blankets, or mats) on a roofing system (a system consisting of a deck and roofing or waterproofing membrane).

Standard (7)

ASTM International-American Society for Testing and Materials International, 2015, ASTM E2400 Standard Guide for Selection, Installation, and Maintenance of Plants for Green Roof Systems.

ASTM E2400 provides general guidance for the selection of plants to green roof systems. It does not include significant information regarding the topics examined in this research.

Data Sheets from Insurance Company

Data Sheets from Insurance Company (1)

FM Global, 2011, Property Loss Prevention Data Sheets 1-35—Green Roof Systems.

The FM Global Property Loss Prevention Data Sheets addresses several topics regarding green roofs and, being a document produced by an insurance company, it provides all inherent risks (i.e., fire and wind risk) and provides green roof warranty values.

Guidelines

Guidelines (1)

FLL-Forschungsgesellschaft Landschaftsentwicklung Landschaftsbau e.V. – The Landscaping and Landscape Development Research Society e.V., 2018, Guidelines for the Planning, Construction and Maintenance of Green Roofing.

The FLL standard for green roofs is the most extensive code of practice in the world, exclusively dedicated to green roofs. Considering the popularity of English edition it is considered the most reliable guidelines worldwide regarding green roof systems, and it is "The Forefather" to all standards and vegetative systems developed sine the release of the first edition.

The "modern green roof technology," with its multilayer system, was first developed in Germany and then exported in the other European Countries and all around the world. In Germany, the *Guidelines for Planning, Construction and Maintenance of Green* has served as the accepted regulations for green roof design for buildings. Thus, the FLL Green Roof Guideline is internationally known as the most extensive design-guide and has served as a benchmark for other countries developing their own standards. It is the primary point-of-reference for all guidelines produced globally.

Guidelines (2)

CIBSE—Chartered Institution of Building Services Engineers, 2007, Knowledge Series 11 Green Roofs.

The KS11 Green Roofs (CIBSE Knowledge Series 11) is a brief document, organized in a question/answer format. It addressed several topics, including cost considerations.

Guidelines (3)

CIBSE—Chartered Institution of Building Services Engineers, 2013, Guidelines for the Design and Application of Green Roof System.

The CIBSE Guidelines for the Design and Application of Green Roof System are general guidelines to address the design and the installation of green roofs. The document is extensive and practical.

Guidelines (4)

Groundwork South Yorkshire, 2014, The GRO Green Roofs Code.

The GRO—Green Roofs Code is a concise document that summarizes the green roof design, specifications, installation methods, and maintenance.

Guidelines (5)

State of Victoria through the Department of Environment and Primary Industry, 2014, Growing Green Guide. A guide to green roofs, walls and facades in Melbourne and Victoria, Australia.

The Growing Green Guide is an extended technical guideline that provides advice on the factors that need to be considered to design, construct, and maintain green roofs, walls, and façades.

Guidelines (6)

GRHC—Green Roofs for Healthy Cities, 2014, Green Walls 101: Systems Overview and Design.

The Green Walls 101: Systems Overview and Design guideline provides high-quality, technical information regarding the design and maintenance of green façades/walls, specifically vertical greenery.

Manuals

Manual (1)

GRHC—Green Roofs for Healthy Cities, 2014, Green Walls 101: systems overview and design.

This document is considerable due to the fact that there not yet standards and complete guidelines on vertical vegetative systems.

Manual (2)

GRHC—Green Roofs for Healthy Cities, 2011, Advanced Green Roof Maintenance.

It addressed exclusively the topic of the maintenance of green roofs. It presents as a comprehensive and detailed document intended for people responsible for green roofs and the maintenance of the vegetation. It is also relevant for designers and developers. The provided information includes the description of practical operations.

Building Codes

Building Code (1)

International Code Council, IBC-International Building Code 2018.

Building Code (2)

International Code Council, IGCC-International Green Construction Code 2015.

Building Code (3)

International Code Council, IEBC-International Existing Building Code, 2012.

Considering the complexity of these codes, only areas where the content on green roofs and façades/walls, relating to the four topics examined, have been identified. Furthermore, the compatibility of the requirements in existing green roofs with the requirements specified in these building codes was not studied, as this would require an additional, specific research study.

Safety Codes

Safety Code (1)

NFPA—National Fire Protection Association, 2015, 101: Life Safety Code.

Safety Code (2)

NFPA—National Fire Protection Association, 2015, 5000: Building Construction and Safety Code.

Like the building codes, only areas where the content on green roofs and façades/walls, relating to the four topics examined, have been identified, due to the complexity of the documents.

City Codes

City Code (1)

Toronto Municipal Code, 2013, Chapter 492: Green roofs.

The code provides the mandatory provisions for green roofs in Toronto.

City Code (2)

Toronto Green Roof Construction Standard-TGRCS, Supplementary Guidelines.

The Toronto Green Roof Construction Standard provides necessary information and references to other standards for green roof requirements that may not be developed in the Toronto Municipal Code or the Ontario Building Code. This is the only document that has been analyzed that provides additional guidelines, with the view to provide clarification between two existing codes. This is an important example because it shows how a brief document can help ensure that green systems follow the requirements set in place by different documents.

Research Reports

Research Report (1)

Department for Communities and Local Government-London, 2013, Fire Performance of Green Roofs and Walls.

The Fire Performance of Green Roofs and Walls reviews existing documents on the fire performance of green roofs and walls and shows results of tests that have been performed on green systems. Existing documents that have been reviewed in this document have also been analyzed in this research study: FLL, ANSI/SPRI VF-1, FM Global Data Sheets, and The GRO—Green Roofs Code. It also analyzes fire prevention requirements in building regulations and determines if the standards for green roofs and façades/walls comply with these. One limitation of the research report by the Department for Communities and Local Government, London, is that it is not complete although it provides the results of the fire tests on green roofs and walls and important fire safety information, it does not include the testing methodology and other references are missing.

Research Report (2)

National Association of State Fire Marshals: Tidwell, Jim and Jack J.Murphy, 2010, Bridging the Gap: Fire Safety and Green Buildings. A Fire and Building Safety Guide to Green Construction, 2010.

Bridging the Gap: Fire Safety and Green Buildings is a research report that analyzes different green building construction elements in terms of fire safety, including brief descriptions of fire safety considerations for vegetative roof systems.

Research Report (3)

FPRF—Fire Protection Research Foundation, 2012, Fire Safety Challenges of Green Buildings: Final Report.

The Fire Safety Challenges of Green Buildings: Final Report addresses fire risks and fire hazards in green building design. It identifies past fire accidents, provides fire risk and mitigation approaches for several green building elements, and assesses four green codes (i.e., LEED, BREEAM, GREENMARK, and IgCC) for possible fire safety concerns that may not be considered.

CHAPTER 4

Standards of vegetative roofs

Standard (1): UNI—Ente Italiano di Normazione, UNI 11235:2015 Istruzioni per la progettazione, l'esecuzione, il controllo e la manutenzione di coperture a verde (tr. Criteria for design, execution, testing, and maintenance of roof gardens)[1]

UNI 11235 defines general standards and requirements for green roofs applied in Italy (it means specifically adaptable to Mediterranean climates) and provides complete technical information to develop professional agreements between clients, planners, and contractors.

UNI 11235:2015 is the second edition of this standard and is an extensive revision and expansion of the first edition, which was published in 2008.

Applicability

- Green roof systems, including intensive, simple-intensive, and extensive greening;
- As a standard that was written for Italy, it provides instructions for green roofs in Mediterranean climates and it is an exhaustive standard for green roofs in warmer climates.

Structure

The standard is divided into an introduction, 11 chapters, and 3 appendixes:

Introduction: Description of green roof performance, divided into three main categories: urban design, ecological benefits, and economical/environmental benefits

1. Synthesis of the purpose and field of application
2. List of all quoted standards

[1] The translation of the standard excerpts is by the author.

Green Roofs, Facades, and Vegetative Systems
ISBN 978-0-12-817694-8
https://doi.org/10.1016/B978-0-12-817694-8.00004-2

3. List of terms and definitions, specifically for agronomic words
4. List of agents and requirements of the green roof
5. Design instructions (this is the most extensive chapter because it includes requirements and designs for every layer of green roof systems)
6. Definition of layering and the classification of the green roof systems
7. Explanation of the design documents
8. Lists of materials and components of green roofs
9. Instructions for construction and installation of materials and components
10. Instructions for controls and testing
11. Maintenance and maintenance programming explanations

Appendix:

A. Instructions for the substrate design

B. Lists of all vegetation typologies for green roofs

C. Instructions for measuring green roof performance tests (control of acoustic insulation, energy flux, runoff coefficient, etc.)

This standard is based on a performance-oriented design approach (performance of each layer and the entire system is widely specified) and it has a high level of completeness, with very detailed technical information for system and layering characteristics, and synthesis of information for all participants (i.e., designer, contractor, maintenance worker, client/developer, and the producer/manufacturer). There are also detailed drawings of the green roof layering.

The most significant limitation is that this standard is only released in Italian, unlike the other analyzed documents, which are all in English.

About

Fire risk/Fire safety

N/A.

The topic is not specifically addressed in any section.

Structural safety/Wind design

The requirements to protect green roofs from wind forces are explained in §5 **Instructions for designing (pp. 6-38)** and are summarized in the following table:

Section	Topic	Text
§5.2 **(p. 7)**	Anchoring systems	In §**5.2 Analyses of the context (p. 7)** some wind precautions are explained: Prevailing winds produce strong stress on the plant species. Therefore, the following characteristics must be evaluated: characteristic of the foliage, height of the vegetation,

Section	Topic	Text
		anchoring capacity of the root system, the stem, and branches elasticity. Then, it is important to evaluate the impact of surroundings buildings. As last, temporary or permanent anchoring devices should be considered in areas with specific wind conditions.
§5.4.9 **(p. 34)**	Wind bracing system	The anchorage of the vegetation may be temporary or permanent depending on the wind conditions, the radical apparatus intrinsic anchor, and the elasticity of the stem. The angle between bracing and the ground must not be greater than 60 ° when they are employed to anchor the stem systems. The design of anchorages is carried out considering the action of the wind on the surface exposed to the same, consisting of the plant species in question. The action of the wind, with an amplifying factor of at least 1.5, must be countered, by gravity, from the element anchoring or from all of these, each for their own share.
§5.4.10 **(p. 34)**	Antierosion layer	For roofs with inclination greater than 15°, the adoption of such layer that, in general, can be constituted by geomats, biomats, biotextiles, geocells.
§5.4.13.3 **(p. 37)**	Use of ballasts	In extensive shell, in particular in the perimeter zones exposed to the wind suction depression, they should be arranged bands of ballasting of inert material, of a minimum width of 50 cm. In general, in correspondence of protruding bodies (perimetric flaps, skylights supports, etc.) Can be positioned to service bands and protection of at least 50 cm wide. This value can still be reduced by 30% for roofs with parapet heights of at least 50 cm. On the perimeter, attention must be paid to the wind load as defined by the current regulations and in the UNI EN 1991-1-4 and consequently evaluating or possibly an increase of the thickness of the crop layer, or the insertion of ballasting groups. Materials should not be placed above the crop layer but above the drainage layer or

Continued

Section	Topic	Text
		protection. In case of use of gravel, this must be round and washed.
§5.4.12.2 **(pp. 35–36)**	Climatic context in the design of vegetation	The wind effect on the water retention capacity of the substrate, the reflected radiation sensitivity, and heat accumulation in the substrate require the choice of plant species with higher resistance to drought. Furthermore, the composition of the plant associations and their spatial distribution must aim to reduce the wind effect on the vegetative layer, also thanks to passive elements, extraneous to the same layer.

§**5.4.4.3 Designing the waterproofing membrane (pp. 25–28)** and §**5.4.4 Designing the membrane against root penetration (pp. 28–30)** are dedicated to waterproofing membranes and the membranes preventing root penetration. Other standards are quoted in this section and installation instructions are provided.

§**Appendix B.5 Trees (pp. 62–63)** provides suggestions for the trees selection:

> "The main problem that presents the use of trees on green roofs is linked to the vertical tree stability, due not only to the limited depth of the substrate layer, but also the possibility given to the roots of expanding laterally and radially, increasing the anchoring surface. Today the nursery offers plants grown in 'flat root-ball' more suitable for use in systems of cultural layers of reduced thickness. For taller plants, appropriate building anchoring systems must be provided.
>
> There are species that should be avoided on green roof applications, because they are characterized by the development of aggressive rhizomes or very vigorous root apparatus that require very deep cultural layers and high nutrient availability.
>
> The following is a list of plant species to be avoided:

- All species of bamboo: Arundinaria fargesii;
- Fargesia murielae, Arundinaria murielae;
- Fargesia nitida, Sinarundinaria nitida;
- Phyllostachys sp., Pleioblastus aleosus;
- Pleioblastus pumilus: Pseudosasa japonica;
- Sinarundina fastuosa;
- Reed: Miscanthus floridus;

- Miscanthus sacchariflorus, Miscanthus sinensis;
- Big/aggressive grass graminacee;
- Arundo donax, Carex glauca;
- Leymus racemosus, Phragmites australis;
- Spartina pectinata;
- Shrubs: Amelanchier sp., Clethra alnifolia;
- Gaultheria shallon;
- Hippophae rhamnoides;
- Sambucus nigra, Rhamnus frangula;
- Buddleia davidii, Polygonum sp.;
- Trees: Salicone, Salix caprea;
- Salix babylonica;
- Populus alba; Populus nigra; Populux X, Ailanthus altissima;
- Taxodium distichum."

Irrigation

§**5.4.14 Designing the irrigation (pp. 37–38)** provides irrigation standards for traditional gardens: "For the design of green roofs it is necessary to identify the water needs of vegetation and size the different types of system to the requirements.

The main used systems are:

- Sprinkler irrigation (from above) or by sprinkling;
- Drip irrigation (ground);
- Sub-irrigation (from below).

They must be chosen considering the specific site conditions (altitude, winds effect).As regards the sub-irrigation, it is needed a specific design considering the characteristics of water accumulation layer."

§**5.3.4.1 Descriptive parameters of the system (p. 11)** recommends an irrigation system installed in all types of green roofs (including extensive green roofs) for guarantying water during arid weeks. An available water supply (obligatory in all green roofs) also represents a security factor in case of a fire.

Maintenance/Cost estimation

§**5.3.5 Level of vegetation maintenance (pp. 13–14)** provides general information about maintenance.

The maintenance levels of green roofs (maintenance requirements) are defined according to the classification of green roofs (i.e., extensive or intensive) and quantified by the amount of man-hours of maintenance are required each year on a square meter basis (Tables 2 and 3).

Table 2 Example of maintenance effort for equal surface (p. 13).

Green typology	Time of work/year
Extensive with minimum maintenance	3–4 h/year
Extensive with low maintenance	6–7 h/year
Intensive with low maintenance	13–14 h/year
Intensive with medium maintenance	About 25 h/year
Intensive with high maintenance	About 30 h/year

Table 3 Classification of green roofs according to maintenance (p. 13).

Level	Man-hours/m^2/year	Green typology
1	mh $<$ 2 min	Extensive with minimum maintenance
2	3 min $\geq$ mh $>$ 4 min	Extensive with low maintenance
3	4 min $\geq$ mh $>$ 8 min	Intensive with low maintenance
4	8 min $\geq$ mh $>$ 15 min	Intensive with medium maintenance
5	mh $>$ 15 min	Intensive with high maintenance

The maintenance requirements are defined by identifying the following:

"The maintenance should be defined at the design stage. It determines the cost for ownership and is closely related to the sustainability of the project from the economic and environmental point of view.

The maintenance levels are defined as following:

Levels 1 and 2—extensive greening:

- Maintenance operations are limited to the control of the systems.
- For vegetation, maintenance consists in monitoring the physiological and health state, the possible presence of parasites, and weeds that may affect the functionality of the system.
- Irrigation can be occasionally needed to avoid water stress for plants.

Levels 3, 4, and 5—intensive greening:

- Maintenance operations, in addition to the control of vegetation and systems, include all the agronomic activities required for green areas.

Irrigation is necessary for the preservation of the intended performance of the green roof and must be the object of a specific design.

The maintenance of intensive green roofs is proportional to: needed water of the system, maintenance operations (man-hours/year) and all the used energy. It is influenced by all the elements of the context, of the site and of the roof, as well as by the architectural and landscape design choices."

All Maintenance is extensively reviewed in §**11 Maintenance (pp. 49-53)**:

This section emphasizes the need for a maintenance plan and categorizes and describes each type of plan:

"For green roofs, a maintenance plan is expected according to the standard UNI 11540:2014. The maintenance plan is the document that provides plans and programs the maintenance of a green roof in order to maintain the functionality over time, the quality features, efficiency, and economic value. The maintenance plan has different contents according to the characteristics and the functions of the project and consists of the following documents:

(1) The user manual;
(2) The maintenance manual;
(3) The maintenance program."

The maintenance plan is then synthesized into four different categories:

(a) Maintenance of greening (plants);
(b) Maintenance of the drainage system;
(c) Maintenance of the gutters, drainpipes, and the waterproofing membrane;
(d) Maintenance of the systems: irrigation and electrical.

Type of maintenance		Description
(a) Maintenance of greening	(a.1) Starting maintenance to control	The agronomic operations necessary in the starting phase and all ordinary agronomic operations
	(a.2) Starting maintenance for ordinary maintenance	The operations are usually the same as for the starting maintenance of greening but with a different intensity and frequency
	(a.3) Ordinary maintenance	The ordinary maintenance includes a list of regular operations: irrigation, fertilization, grubbing weeds, cutting grass, pruning, and phytosanitary treatments
	(a.4) Extraordinary maintenance	Operations carried out as results of special situations like intense adverse meteorological events
(b) Maintenance of the drainage system		Annually and before winter, inspection of the drainage system, to eliminate materials, foliage, etc.

Continued

Type of maintenance	Description
(c) Maintenance of gutters, drainpipes, and waterproofing membrane	Early and before winter it is necessary to make an inspection of the gutters and drainpipes removing foliage and other materials
(d) Maintenance of systems: irrigation and electrical	All the system must be maintained according to the reference standards. Several instructions and tips are provided for the irrigation system

The types of maintenance are further explained in the following subsections:

§**11.2.1 Starting maintenance to control (pp. 49–50)**:

(a) The agronomic operations necessary in the starting phase, depending on the type of greening, on the site conditions and the climatic conditions:
- Control of the crop thickness and subsequent possible integration;
- Control of the compaction as a result of frost with possible operation of aeration and integration;
- Tamping of herbaceous plants and shrubs;
- Reseeding failed areas;
- Replanting failed areas;
- Check the efficiency of the retaining systems with possible reinstatement/substitution;
- Control of the irrigation efficiency;
- Control and cleaning of the irrigation elements (shafts, filters, distribution pipes).

(b) All ordinary agronomic operations:
- Irrigation;
- Fertilization;
- Elimination of weeds and dead plants;
- Mowing the turfs;
- Mowing of grassland associations;
- Pruning;
- Pesticide treatments.

Figure 11—Example of maintenance plan (p. 50)

Yearly program of starting maintenance	Extensive green roof—second year											
Type of intervention	**January**	**February**	**March**	**April**	**May**	**June**	**July**	**August**	**September**	**October**	**November**	**December**
Cutting of vegetation as "wild field"												
Check of the irrigation functioning			20				20					
Fertilization of lawns												
Fertilization of perennial herbaceous species												
Fertilization of shrub species												
Phytosanitary treatments Control of substrate thickness			20									
Tamping of herbaceous plants, shrubs, and trees												
Reseeding plant lacks												
Control, with possible reinstatement/substitution, of the anchoring systems												
Irrigation control			10	15	15	20	30	30	15	10		
Total minutes maintenance/year			40	15	15	20	50	30	15	10		

§11.2.2 Starting maintenance for ordinary maintenance (p. 50)
"The operations that constitute the starting maintenance for ordinary maintenance include all the works usually necessary to reach the ordinary maintenance, but with a different intensity and frequency."
§11.2.3 Ordinary maintenance (p. 51)
"The purpose of ordinary maintenance is the maintenance over time of the greening according to the expected functions, by the agronomic operations with according to the type of greening, the site conditions and the climatic conditions.

Such operations have the same purposes as similar works on ordinary green (green and gardens on ground level), but must be implemented by appropriate methods to the roof as not to prejudice the stratigraphy and the function of the waterproofing membrane.

In particular, it's important to avoid overloads by machines and/or materials, impacts and use of pointy or sharp equipment not to appropriate for green roofs.

The ordinary maintenance includes all the following operations:

- Irrigation;
- Fertilization;
- Elimination of weeds and dead plants;
- Smoothing of turf;
- Mowing the turfs;
- Mowing of grassland associations;
- Pruning for/aesthetic functional purposes;
- Pesticide treatments."

§11.2.3 Extraordinary maintenance (p. 51)

"Extraordinary maintenance is carried out during special situations that may occur during the life of the green roof such as, for example, adverse meteorological events, extraordinary plant diseases, etc."

§11.3 Maintenance of the drainage system (p. 51)

"Annually and before winter, it must be carried out an inspection of the drainage system (where possible), to eliminate materials, leaves and other plant parts that may reduce the efficiency of the drainage system."

§11.4 Maintenance of gutters, drainpipes and waterproofing membrane (p. 51)

"It is necessary, annually and before winter, to make an inspection of the gutters and drainpipes terminals with their possible cleaning in order to avoid occlusions.

The accessible parts of the waterproofing membrane must be cleaned."

§11.5 Maintenance of the systems: irrigation and electrical (p. 52)

"The procedures for maintaining the irrigation and electrical systems must be applied according to the reference standards. For what concerns the irrigation system, it must be always cleaned, efficient and in perfect functioning."

§11.5.1 Programmed periodic interventions on irrigation systems (p. 52) provides further specification for the correct irrigation working and instructions are provided:

The operations are differentiated according to the different systems adopted (sub-irrigation, rain, to drip surface, etc.).

As a rule, the following actions must be performed:

- To fill the system in late winter and to activate the system;
- Periodic inspections during the growing season to verify the correct working;
- To empty and shutdown the system in the early autumn.

The interventions include:

- To verify the correct operation and the possible modification of the irrigation frequency of the programming unit;
- To verify the efficiency of the valves, joints and driplines.

Also, in order to prevent that the inevitable impurities occlude the irrigation nozzles, at least once a year it is necessary to empty the entire system, zone by zone.

§11.5.3 Further instructions provides more rules to guarantee the success of the system. This section and the subsections provide further instructions and tips for a correct working of the irrigation system, in particular specific rules are provided for protection by winter frost:

- Closing the water source of the irrigation system;
- Insulating the principal valve;
- Insulating the irrigation system with insulating foam;
- To empty the irrigation pipes.

Standard (2): MCCAA—Malta Competition and Consumers Affairs Authority, Standards and Metrology Institute, SM 3700:2017 Green Roofs—Criteria for the planning, construction, control and maintenance of Green Roofs.

The Maltese standard for green roofs sets principles and requirements for designing, constructing, and maintaining green roofs. This standard is a similar document, for conception, to the Italian UNI 11235:2015 it provides full requirements for realizing a green roof and specific reference standards where needed. Also this standard is self-sufficient for design, construction, and maintenance of a green roof, so it is a highly qualified document for all the professionals and trade people in the field of green roofs.

Applicability

- Green roof systems, including intensive, simple-intensive, and extensive greening;
- General green roof requirements and specific requirements for each green roof layer;
- Necessary technical information to develop professional agreements between clients, planners, contractors, and maintainers.

Structure

The standard is divided into 10 chapters and 3 appendixes:

1. Scope of the standard
2. Normative references
3. Terms and definitions
4. Introduction
5. Types of green roofs
6. Planning of green roofs
7. Materials and components
8. Other considerations
9. Drainage requirements
10. Controls (final inspection)
11. Maintenance

Appendix A—Determining the runoff reference value/coefficient of discharge

Appendix B—Determination of compressibility and permeability

Appendix C—Reference figures

This standard is recognized by its high level of completeness, including very detailed technical information for green roof systems as a whole and for each individual component and layer. Furthermore, as a standard, it serves as a guarantee of the quality and success of the system for the client and all involved in the design, construction, maintenance, development, and production/manufacturing.

About
Fire risk/Fire safety

The fire topic is mentioned in §**6.2 Structural Requirements (p. 11)**: above others, the "Structural requirements in relation to Green Roofs mainly refer to:

- [...]
- fire characteristics
- [...]"

In §**6.4 Site Analysis (p. 12)** it is mentioned that: "Particular attention must be given not only to exposure to sun, but also to any reflective surfaces such as glazed elevations and walls." This recommendation is referred just to plants health and not to fire risk.

In §**7.4.1 Components (p. 23)** of §**7.4 Vegetation support layer** about the composition of the substrate it is mentioned that the vegetation support layer "shall have all the requisite basic physical, chemical and biological properties:

[...]

vi. sufficient presence of organic matter to permit the formation of a stable ecosystem, but not too high in order to avoid excessive structural alterations and reduce any fire hazards;

[...]"

§**7.4.7 Organic matter (p. 27)** content specifies the maximum quantity of organic matter in substrate, which is fixed in:

- for intensive green roofs: $\leq$ 80 g/L
- for extensive green roofs: $\leq$ 60 g/L

In §**7.6.7 Fire characteristics (p. 36)** under §7.6 Drainage course **(p. 34)**, it is specified that the used materials shall not be in flammable materials or pose fire hazard: "Both granular and geosynthetic drainage layers shall not be composed of flammable materials or pose as a fire hazard.

In general, the requirements for aggregates used for drainage layers apply to the material after it has been compacted to the defined laboratory standard.

The different properties associated with the materials need to be assessed against the conditions which apply to the position and site where they are to be used, in order to ensure that they are suitable."

In **Table 11 Evidence is required for the properties of materials used in granular drainage courses to verify suitability of materials (p. 39)**, among several aggregate properties of the substrate, it is mentioned that the fire characteristics should be "E."

In §**8.7 Fire characteristics (p. 55)** it is declared that: "Due to the island's physical factors, most of the plants which are ideal for use on a Green Roof tend to be woody. These could pose a low to medium risk of fire threat. Certain species such as *Jacobea maritima* are considered as good fire breakers. Other ways of reducing fire risk include:

- boarder of pebbles, gravel or pavers which allows at least 50 cm between the vegetation area and any openings in the roof (skylights, windows) or any vertical elements such as a wall with windows;
- 30 cm high barrier constructed of non-flammable material every 40 m;
- 1 m wide strip of solid pavers, gravel or pebbles;
- an emergency sprinkler system used as a first aid fire-fighting measure."

This statement—inside the standard—is an important tip for the designers to consider the fire risk as part of the design phase.

Structural safety/Wind design

The wind topic is mentioned in several sections:

In §**6.1 Agents and requirements (p. 11):** above others, the agents that "shall be taken into consideration during the planning of Green Roofs are:

- [...]
- Live and Dead loading
- Static and Dynamic Loading
- [...]"

In §**6.2 Structural Requirements (p. 11)**: above others, the "Structural requirements in relation to Green Roofs mainly refer to:

- [...]
- protection against lifting or excessive drag forces caused by the wind
- protection against slipping and shearing
- tree supports
- [...]"

In §**6.4 Site Analysis (p. 12)** wind is mentioned as an important factor to consider in the site analysis: "Wind produces significant stress to plant species. [...] Adjacent building and structures would also need to be considered when analyzing the effects of wind, as they can create eddies and vortices which could affect plant survival and development. As a result, it would be opportunistic to consider the need for permanent or temporary anchoring systems."

In §**6.5 Certification of the structure (p. 13)** it is written that a certification by a qualified engineer is mandatory before planning, design and construction phases: "Prior to the commencement of the planning, designing and constructing stages of the Green Roof, the building and roof in question shall be certified suitable to withstand the loading imposed by the Green Roof and other related engineering aspects in connection to the Green Roof. This certificate shall be issued by a suitably qualified engineer.

Structural alterations/redesigning might be required for the building and/or roof to be able to withstand the Green Roof. Restrictions might arise in the installation of certain types of Green Roofs."

§**6.10 Design loads (p. 15)** specifies the typical design loads for different plant types like Sedum, grass, shrubs, trees according to the vegetation high.

§**8.6 Wind loads (p. 54)** addresses the issue of wind loads in general terms, referring to other standards:

- DIN 1055-4 "Einwirkungen auf Tragwerke — Teil 4: Windlasten' - Action on structures - Part 4: Wind loads";
- DIN 1055-100 "Einwirkungen auf Tragwerke — Teil 100: Grundlagen der Tragwerksplanung — Sicherheitskonzept und Bemessungsregeln' - Actions on structures - Part 100: Basis of design, safety concept and design rules."

The section cites: "Wind forces can act on structures with positive and negative pressure and through friction, due to both friction and pressure drag. In some cases, the wind results in lifting forces which can be dangerous in case of insufficient anchorage of certain elements or structures. The strength of these forces is a direct function of wind strength speed and direction and of the shape and height of the building in question. The building form itself has an important impact on the resulting flow speed encountered at various points around the building. On the roof, a separation layer of highly turbulent flow will occur. This layer can have a height which extends up to a few meters above roof level. This means that most of the components making up the Green Roof will be located in this separation zone and therefore in a region of highly turbulent flow. The resulting dynamic wind loads can damage anything built on top of the roof, either during construction or after work has been completed. Action therefore needs to be taken at the planning stage to prevent waterproof linings, and other materials for the structure of the roof, from being lifted off by the wind.

Where waterproof linings/root-penetration barriers are not affixed rigidly, the layered structure of the Green Roofs must be used to prevent the membranes/linings from being lifted by the wind. During the construction stage any lose materials should be weighed down to prevent lifting by wind.

On Green Roofs, every effort is usually made to ensure that loads and the depth of the layered structure are kept to an absolute minimum. There are situations in which this depth needs to be increased or heavier materials need to be used in order to secure edges and corner areas which are particularly at risk. The critical factor here is the dry load of the layered structure. In certain cases, gravel or slabs may have to be used in order to secure the roof edges and corners.

The calculation of the loads necessary to protect against wind suction can be done following recognized standards (e.g., DIN 1055-4 in combination with the coefficient of wind action in DIN 1055-100). Data over the securing of waterproofing on roofs against wind action can be found in the ZVDH/HDB—"Rules for Roofs with Waterproofing." More detailed design of the Green Roof may consider the adoption of Computational Fluid Dynamics (CFD) simulation tools, in order to predict the air flows occurring over the roof which can then be used to determine the resulting loads. To perform such CFD simulations, the guideline document "Best Practice Guideline for the CFD Simulation of Flows in the Urban Environment" by Franke et al. should be used.

These requirements, in respect to gravel or concrete slabs, relate solely to the vertical load. This takes no account of the following factors which also apply to Green Roofs:

- the coarseness of the vegetation;
- the load generated by residual moisture in the soil;
- the load generated by the vegetation;
- the bonding of the layers through the action of the roots in comparison to loose material;
- the flow permeability of the vegetation support layer, which diffuses the pressure differential between the top and bottom of the vegetation layer, thereby reducing the load.

The listed criteria will lead to a reduction of the wind load. For the calculation, DIN 1055-4 prescribes an aerodynamic coefficient for the outside pressure of CPE 10; but as stated, more accurate flow simulations are strongly recommended."

At last, requirements referred to structural safety and wind design for furnishing are written in the §**8.10 Furnishing (p. 57)**:

§8.10.1 Types (p. 57)

"Furnishings include items such as:

- trellises;
- pergolas;
- lighting;

- ponds;
- PV panels and solar water heaters.

The layout and installation of furnishings is site specific and special consideration will need to be made to the construction, structural and architectural requirements."

§8.10.2 Requirements (p. 67)

Furnishings must be sturdy and set up and secured in such a manner as to spread their weight evenly. It is particularly important to ensure that no stresses are generated on the supporting base. Allowance must be made for point and/or surface loads and for wind loads."

§8.10.3 Installation (p. 57)

"Furnishings may be installed by means of:

- a system which anchors them to the roof and distributes the load, using mountings incorporated into the design of the roof; or
- flat or truss-type foundations.

Provision for furnishings can be incorporated into the structural design by means of mountings which protrude above the waterproof lining. Here, attention must be paid not only to the requirements in respect of static loadings, but also to the instructions relating to roof penetrations.

Installation of furnishings which do not form part of the original roof design should only be undertaken in exceptional circumstances. Where this does happen, care must be taken to ensure that the continuity of the waterproof lining/root-resistant membrane, the thermal insulation layer and waterproofing is not disturbed. The construction of spread or truss-type foundations means that anti-bonding and protective layers will have to be installed above the underlying layers of the roof structure.

The subsequent necessity for roof penetrations, to secure the furnishings which have been planned, should be avoided. Should foundations for anchoring the furnishings still be necessary, then alternative methods should be considered, such as securing pergolas, trellises, lamps or benches to spread or grid foundations where possible."

Irrigation

§6.13 Irrigation (p. 17) mentions general requirements for green roofs: "The irrigation requires at least one water connection at roof level. The number of pipes, connections and the necessary water pressure will depend upon the location and structure of the building. The size, layout and species used will also determine the overall layout of the irrigation system. The design of the irrigation system should be factored in during the planning stages of the Green Roof. If possible rain water collected in a cistern is used to irrigate the green roof plants."

Then, §**8.1.3 Additional irrigation (p. 50)** under the §**8.1 Water retention (p. 48)** adds information about irrigation, which is not considered as obligatory but just recommended in regions with low rain: "In regions of low precipitation, in addition to water stored in the drainage layer, watering may be required in all types of Green Roofs. Consideration shall be given to the watering requirements of the individual species. However, efforts shall be made to use xerophytic plants.

Additional watering may be provided by using:

- a hose
- hose and sprinkler
- spray hoses
- driplines
- an overhead irrigation system
- automated watering systems with an in-built reservoir

Where sprinklers, spray hoses or driplines are used, the system can either be operated manually or controlled by means of a timer. A hand-held hose will have to be used to water any areas untouched by a sprinkler system (edges and corners), and other partial areas which are tucked away under roofs, affected by the deflection of the water jet, or affected by the wind.

Irrigation systems, installed above or beneath the ground, may be operated manually, by a timer or they may be fully automatic. The lines must be corrosion- and frost-proof, or be fitted with a facility for draining the system completely to avoid frost damage.

A reservoir-based watering system may be fitted with an automatic or semi-automatic water feed. Precipitation is stored in the drainage course and acts as a reservoir, the capacity of which will depend on the type and depth of the drainage course involved. A minimum clearance needs to be maintained between the peak level in the reservoir and the filter layer, in order to prevent the vegetation support layer from becoming waterlogged.

Soft landscaping must not span across expansion joints and at any junction between walls and floors (e.g., along the parapet walls) and where unrestricted access must be available at all times.

The management of the irrigation system may be carried out using soil moisture sensors and water counters to control the amount of water applied to the roof. The water storage properties of the system must be taken into consideration so as to avoid increased water runoff from the system and possible damage to the plants."

Maintenance/Cost estimation

§**6.6 Access and use of Green Roofs (p. 13)** premises the topic of green roofs accessibility: "Areas designated for use by people shall be safe to use and access. Footpaths and paved areas shall be provided. Turfed (grassed) areas purposely planted for human traffic is acceptable, although consideration must be given to elevated use of resource and maintenance requirements.

Where Green Roofs need to be accessed for works as defined by Act XXVII/2000, compliance with obligations as outlined in said Act need to be ensured.

Planted areas are generally not suitable for public use. In extensive Green Roofs, access is normally restricted to maintenance and servicing."

The standard does not provide more information on green roofs access.

§**10 Maintenance (pp. 60–62)** deepens the topic.

The first aspect to underline is that the standard imposes a maintenance plan: "The maintenance plan is a document which foresees, plans and schedules maintenance works, so as to preserve over time the functionality, quality, and efficiency of the Green Roof system.

The maintenance plan assumes different content depending on the importance and the intervention specificity, and consists of the following operational documents:

- the user manual;
- the maintenance manual;
- the maintenance program."

§**10.1 General (p. 60)** classifies the maintenance according to the layer/system to be maintained: "The maintenance regime can be categorized as follows:

- Maintenance of the soft landscaped areas;
- Maintenance of the drainage system;
- Maintenance of the water collection system, drains and damp proof membrane;
- Maintenance of services."

The maintenance is then classified in three types:

(1) §**10.3 Initial maintenance to achieve the projected greening (pp. 60–61)**

"Such maintenance works shall include what is necessary to achieve the planned outcome of the Green Roof, including the provision and appropriate measures to protect the substrate and vegetation from erosion and other physical anomalies.

Initial maintenance shall include:

a. horticultural operations required in the initial stage: this varies depending on the type of greening, microclimate and physical environment including:
   - Confirming the depth of the substrate with the possibility of adding more substrate to attain the original depth;
   - Planting further herbaceous plants, ground cover and shrubs to fill in any gaps;
   - Reseeding failed areas (ground cover established through seeding);
   - Replanting failed areas (ground cover with planting of trees and shrubs);
   - Checking the effectiveness of anchors and adjusting if necessary;
   - Checking the effectiveness of the irrigation system;
   - Checking and servicing all components of the irrigation system including filters, solenoid valves, joints, tubing, drippers, etc.

b. The required maintenance works shall differ, depending on the type of roof greening, microclimate, and season. These include:
   - irrigation;
   - fertilization (kept to a minimum);
   - weeding control;
   - grass cutting;
   - pruning;
   - plant protection practices."

(2) §10.4. Long-term maintenance (p. 61)

"Long-term maintenance is a continuation of the initial maintenance. The scope of long-term maintenance is to preserve the characteristics of the initial design scheme over time, without undermining the efficiency and proper performance of the Green Roof system.

It shall be imperative to avoid impact loading, overloading the roof with the use of machinery and/or materials, the use of pointed and sharp equipment which could in any way damage the layers below the growing medium.

Long-term horticultural maintenance includes all the works required for the survival of the vegetation layer which include:

- Irrigation;
- Fertilization (kept to a minimum);

- Uprooting of unwanted vegetation;
- Grass cutting;
- Pruning diseased and dead stems;
- Pruning for design purposes;
- Plant protection practices."

(3) §10.5. Unscheduled maintenance (p. 61)

Such maintenance shall be performed following unexpected situations during the lifespan of the Green Roof and include:

- adverse meteorological events,
- outbreaks of plant diseases,
- particularly dry spells, or
- other.

Then, the following paragraphs address specific topics:

§10.6 Maintenance to the rain water drainage system (p. 61)

§10.7 Maintenance of services (p. 62)

§10.8 Periodic inspections on the irrigation system (p. 62)

"Maintenance operations shall respect the different irrigation systems adopted (surface drip, sprinklers, etc.). Maintenance operations shall include:

- Confirming adequate water availability before and during the dry season;
- Periodic inspections of the irrigation system during the growing season;
- Shutting down or re-setting the irrigation system during the wet season.

If required, interventions shall include:

- Verifying the correct operation and the possible modification of the irrigation time;
- Programming the control unit during different seasons;
- Verifying that leaks are absent in the irrigation system.

To avoid the inevitable clogging of irrigation nozzles, the system shall be bled and flushed every other year."

§10.9 Repairs (p. 62)

§10.10 The irrigation control system (p. 62)

"It is advisable that the irrigation system is not the pressurized type. All irrigation pipes should not be kept under pressure when the system is not activated. To avoid damage to the pump, it is advisable that if the controller is programmed to control the pump, it would have a cut-off switch to prevent damage to the pump if the solenoid valves do not function."

The maintenance topic is overall informed and deepen.

Standard (3): ASTM International–American Society for Testing and Materials International, 2014, ASTM E2777 Standard Guide for Vegetative (Green) Roof Systems

This standard guide provides technical requirements for extensive and intensive green roof systems (i.e., type of plant, wind scour resistance, soil reinforcement, separation/filter layers, drain layers, water retention layers, protection layers, and root penetration barriers).

Applicability

- A general guide for professionals designing or constructing extensive and intensive green roofs systems, with up to a 15% slope.

Structure

The standard is divided into eight parts and one appendix:

1. Scope of the standard guide
2. Referenced documents
3. Referenced documents for the terminology and definitions
4. How to use this standard to benefit green roof systems
5. Explanation of principles related to the selection of vegetative green roof systems (i.e., design intent and building function, sustainability, design considerations, etc.)
6. Recommendations to ensure quality of green roof system (i.e., performance requirements, submittals, maintenance program, project check list, implementation)
7. Technical requirements for plants, media, wind scour resistance, soil reinforcement, separation/filter layers, drain layers, water retention layers, protection layers, root penetration barriers, and membranes
8. Summary of the keywords

   Appendix:

   A review of the potential contributions a vegetative green roof system can have, in order to achieve the sustainability objectives of the project.

Characteristics

Although just a general guide, it accurately frames what a green roof system is, explaining its function, introducing important precautions for the design, and explaining the significance of maintenance. Also, the appendix provides a well-constructed summary of how a vegetative green roof system can help a project achieve sustainability objectives.

As specified in the title, this is a standard guide: the document provides general information about green roof design and maintenance, but it is not meant to be an in-depth guide; despite this, it is a useful document for designers and architects.

About
Fire risk/Fire safety

§5.3.9 (p. 5) recommends:

(1) The use of succulent plants in regions where brush fires are an identified threat;

(2) Nonvegetated margins;

(3) Firebreaks.

"In regions where brush fires are an identified threat, it is recommended that designs emphasize foliage cover consisting of succulent plants (for example, *Sedum, Senecio, Delosperma, Graptopetalum, Echeveria,* etc.), and the vegetative (green) roof system should be maintained to regularly remove dead or dormant grass and shrubs. Nonvegetated margins, consisting of coarse stone, gravel, concrete pavers, or stone pavers can be used to set back foliage-covered areas from critical surfaces. Specifically, setbacks for plant foliage are recommended in the following situations: 1) walls immediately beneath the sills of operable windows, and 2) adjacent to hatchways, thresholds, and mechanical equipment. Non vegetated setbacks are recommended for boundaries with roofing/waterproofing membrane systems that are not classified Test Methods E108 Class A and from building surfaces constructed using materials that have not been successfully tested in accordance with Test Method E136. For vegetated (green) roof systems that are not rated Class A or B based on Test Methods E108, additional precautions are recommended, including providing breaks in the vegetative (green) roof system that will limit the area of any contiguous foliage-covered roof zone. Breaks may consist of concrete or masonry curbs that are taller than adjacent plant foliage or non-vegetated strips. Non-vegetated strips may consist of either: 1) coarse stone, gravel, concrete pavers, or stone pavers, or 2) Class A roof covering, as determined by Test Methods E108. All vegetative (green) roof systems should be provided with access to hose-bibs faucets, or an irrigation system that can provide sufficient to water to allow the entire vegetative (green) roof system to be thoroughly soaked within an elapsed time of 2 h. Provisions for introducing fire resistance measures shall comply with requirements of Federal, State, Provincial, or local entities with jurisdiction."

Structural safety/Wind design

§**5.3.8 Wind Resistance (p. 5)** provides some general information regarding wind damage along perimeters and corners and leading-edge systems (i.e., gravel, stone margins, etc.). It suggests referencing entities with jurisdiction to determine requirements for leading-edge systems and references ASCE/SEI 7 for ballast requirements based off of wind velocities and uplift pressure.

"Damage by wind is a concern with vegetative (green) roof system installations, particularly along perimeters and corners, at obstructions such as mechanical equipment, and adjacent to penthouse structures. The potential for damage by wind will vary with building height, building geometry, geographic location, and local topography. Probability of wind damage is greatest with high winds immediately after installation and diminishes as the vegetative (green) roof matures. With many vegetative (green) roof systems methods for temporarily protecting the media prior to establishment of a mature plant ground cover may be advisable. This may include mats or mesh fabricated from organic fibers or geosynthetics, tackifying agents, or the installation of pregrown mats or modules. Various permanent stabilized leading-edge systems may be viable for a particular project, including: gravel or stone margins, unit pavers, strapped or bolted pavers, reinforced media layers, and buried ballasts in conjunction with reinforcing geotextiles. The width of stabilized leading-edge systems depends on the local wind environment, which is specific to each building and geographic setting. Measures shall comply with requirements of Federal, State, Provincial, or local entities with jurisdiction. Methodologies for determining ballast requirements often rely on estimates of wind velocities and uplift pressures based on ASCE/SEI 7. The dry weight should be used when evaluating the ballast weight of a vegetative (green) roof system. In some jurisdictions upper limits on basic wind speed (3 s gusts) may apply to inclusion of gravel or stone ballast, due to the risk of these becoming windborne missiles."

§**7.4 Wind Scour Resistance (p. 9)** provides tips for stabilizing the growing media by wind scour: "Some winnowing of fine particles from the surface of vegetative (green) roof systems should be anticipated when establishing media thicknesses. Some vegetative (green) roof media cannot resist wind scour unless stabilized by a uniform vegetative ground cover. Therefore, rapid establishment of ground-covering plants is an essential consideration when addressing the longevity of a vegetative (green) roof. Where protection of exposed media is of concern, techniques for wind stabilization should be introduced. Depending on the severity of the condition, these may include: temporary mats composed of biodegradable/photodegradable materials, permanent polymeric meshes or nets, and tackifying agents. Pregrown mats or linked modules can also be used to create a wind-stabilized surface."

§**7.5 Soil Reinforcement (p. 9)** provides recommendations for soil reinforcement to prevent damage caused by extreme rainfall and wind events: "There are several reasons why soil reinforcement might be introduced as part of a vegetative (green) roof system. Soil reinforcement may be used to prevent damage caused by extreme rainfall and wind events, or to protect areas of a vegetative (green) roof where outwash from downspouts occurs. Also, soil reinforcement can be used to stabilize turf grass that will be subjected to heavy recreational use. The appropriate test for evaluating the usefulness of soil reinforcing layers are Test Method D4595, for geotextiles, and Test Method D6637 and Test Method D5262, for geo-grid."

Irrigation

N/A.

Maintenance/Cost estimation

§**5.3.1 Maintenance (p. 4)** explains how a maintenance manual should be developed. §**5.3.1** recommends that the manual should include instructions for operation of irrigation systems, instructions for proper weeding and fertilization, methods for recognizing and dealing with commonly encountered problems (insect infestations, weed infestations, bare spots, etc.), and instructions for inspecting exposed elements of the roofing/waterproofing membrane system.

"All vegetative (green) roof systems shall be accompanied by a detailed written maintenance procedures manual, provided by the design professional, vegetative (green) roof installation company or system manufacturer. Maintenance manuals should include instructions for operation of irrigation systems, where relevant, and directions for proper weeding and fertilization. These documents should also include methods for recognizing and dealing with commonly encountered problems, including: insect infestations, weed infestations, bare spots, wet spots or areas with perennial surface water ponding. Depending on the vegetative (green) roof system and site conditions, provisions for employing temporary irrigation should also be addressed. Manuals should also include instructions for inspecting exposed elements of the roofing/waterproofing membrane system, most notably the drains. Minimum requirements for site visitations should be provided."

§**5.3.3 Longevity (p. 4)** describes that the longevity of a vegetative green roof can prematurely fail if components or the roofing/waterproofing membrane system degrade or fail, due to neglectful maintenance practices.

§**6.1.3 Maintenance Program (p. 6)** summarizes information about maintenance requirements and responsibilities for the vegetative green roof system installer or system manufacturer, which should be included in the contract document: "Contract documents should be specific concerning the maintenance requirements and responsibilities of the vegetative (green) roof system installer or system manufacturer. For example:

- §**6.1.3.1**: Procedures for leak detection and repair, as necessary, for the roofing/waterproofing membrane.
- §**6.1.3.2**: Requirements for minimum foliage cover prior to acceptance by the owner. Specify remedies if the cover requirement is not satisfied at the end of the designated establishment period.
- §**6.1.3.3**: Requirements for continued performance, including effective drainage, soil thickness, horticultural viability, etc., provided the maintenance program is followed.
- §**6.1.4**: Required longevity for constituent components."

Standard (4): ANSI/SPRI—American National Standard Institution/Single Ply Roofing Industry, 2017, ANSI/SPRI VF—1 External Fire Design Standard for Vegetative Roofs

The ANSI/SPRI VF—1 standard provides a method for designing external fire spread resistance for Vegetative Roofing Systems. It is intended to provide a minimum design and installation reference for those individuals who design, specify, and install Vegetative Roofing Systems.

Applicability

- Required external fire resistance and the minimum design/installation requirements for flat green roofs up to a 2:12 slope;
- To be used in conjunction with the manufacturer's specifications and requirements for the vegetative roofing system.

Structure

The standard is divided in two main parts. The first part is five chapters in length and provides definitions used in the standard, general design considerations, vegetative roofing system requirements, and general instruction about maintenance.

The second part is the commentary on the standard, divided in several subsections, which specifies some of the previous five chapters, just when needed.

Part 1: Standard

1. The introduction that frames the object of the standard
2. Lists of all the definitions used in the standard (i.e., ballast, area divider, fire-barrier, etc.)
3. General design considerations (very brief considerations about roof structure design or evaluation, roof deck waterproofing layer or roof cover requirements, slope, firebreaks, area divider, border zone) with references to the Commentary
4. Vegetative roofing system requirements (very brief descriptions) with the references to the Commentary
5. Brief description of maintenance requirements

Part 2: Commentary

The Commentary consists of explanatory and supplementary material designed to assist designers and local building code committees and regulatory authorities in applying the requirements of the preceding standard.

The Commentary is intended to create an understanding of the requirements through brief explanations of the reasoning employed in arriving at them.

Characteristics

The standard provides a method to give the minimum requirements for fire design and is consistent with other standards and codes regarding green roofs. It also provides simple and effective instructions for hydrants and maintenance requirements.

About

Fire risk/Fire safety

The standard defines basic requirements and values/measures for the external fire design of green roofs, including requirements.

Among the definition in §**2.0 Definitions (p. 2),** the following are particularly relevant:

§**2.3 Border Zone (p. 2)**

"The region around the edge of the vegetative plantings, where no vegetation exists. It is frequently the perimeter of the roof area, and areas around Penetrations and drains. See **Commentary C 2.3**."

§2.5 Fire Barrier (p. 2)

"A fire-resistance-rated wall assembly of materials designed to restrict the spread of fire in which continuity is maintained."

§2.6 Firebreak (p. 2)

A Firebreak is a section of the roof that is covered with stone Ballast or concrete pavers and acts to slow or stop the progress of a rooftop fire.

§2.10 Penetration (p. 2)

An object that passes through the roof structure and rises above the roof deck/surface. Penetrations consist of, but are not limited to, mechanical buildings, penthouses, ducts, pipes, expansion joints and skylights. See **Commentary C 2.10**.

Among §**3.0 General Design Considerations** (p.3), the following are particularly relevant:

§3.4 Firebreaks (p. 3)

"Where required Firebreaks shall be installed to provide a minimum 6-ft wide (1.8 m) continuous border."

§3.5 Area Divider (p. 3)

"Where required an Area Divider shall be installed to provide a minimum 13-ft wide (4 m) separation zone."

§3.6 Border Zone (p. 3)

"A minimum 3-ft wide (1 m) continuous border free of vegetation and Growing Media."

§4.0 Vegetative Roofing System Requirements (pp. 3–4) is a relevant section:

§4.1 (p. 3)

"The waterproofing system below the vegetation shall be tested per ASTM E108 and meet the fire classification requirements of the authority having jurisdiction."

§4.2 Fire Protection for Roof Top Structures, Joints and Penetrations (p. 3)

"A Border Zone (See **Section 3.6**) shall be provided where Vegetative Roofing Systems abut Non-Combustible rooftop structures, or joints and Penetrations. See **Commentary C 4.2**."

§4.3 Spread of Fire, Protection for Large Area Roofs (p. 3)

"An Area Divider as described in Section 3.5 shall be used to partition the roof area into sections not exceeding 15,625 ft2 (1,450 m2), with each section having no dimension greater than 125 ft (39 m). See **Commentary C 4.3**."

§4.4 Spread of Fire, Protection for combustible features that are part of the green roof design, but not part of the building structure (p. 3)

"An Area Divider shall be installed around combustible features that are part of the vegetative roof design. See **Commentary C 4.4**."

§4.5 Fire Hydrants (p. 3)

"Access to one or more fire hydrants or stand pipes shall be provided."

§4.6 Firebreak (p. 4)

"Firebreaks (See **Section 3.4**) are required where Vegetative Roofing Systems abut combustible vertical surfaces and when terminating at a Fire Barrier."

Among §**Commentary to VF—1 (pp. 5—9)** the following are particularly relevant:

§C2.3 Border Zone (pp. 5—6)

"For design and installation purposes, the roof surface is divided into the following areas: Corners: The space between intersecting walls forming an angle greater than 45 degrees but less than 135 degrees.

Corner Areas: The corner area is defined as the roof section with sides equal to 40% of the building height. The minimum length of a corner is 8.5 ft (2.6 m).

Perimeter: The perimeter area has different size definitions depending upon the method of installing the roof. For Ballasted roof systems the perimeter area is defined as the outer boundary with a width measurement equal to 40% of the building height, but not less than 8.5 ft (2.6 m). For adhered roof systems it is defined as the outer boundary of the roof width measurement equal to the least of the following measurements; 0.1 x the building width or 0.4 x the building width. The minimum width is 4 ft (1.2 m).

The perimeter area for a ballasted roof is larger due to a concern for Ballast blow-off. Field: The field of the roof is defined as that portion of the roof surface, which is not included in the corner or the perimeter areas as defined above."

§C4.0 Vegetative Roof System Requirements (p. 8)

"Effective with the 2018 Edition of the International Building Code, Vegetative Roofing Systems will be required to meet the same fire classification requirements as the roof covering and roof assembly. Due to the many variables (including plant type, plant condition, depth of Growing Media, combustibility of roofing assembly materials, and installation details) and the lack of sufficient experience and test data, classification of exterior fire exposure cannot be made with certainty at the present time. This standard requires that the roof system installed below the Vegetative Roofing System meet the fire classification requirements of the authority having jurisdiction. The standard then uses Border Zones and Firebreaks to protect roof top structures, Penetrations and joints that may be on the roof. It also uses roof divider areas consisting of ASTM E108 Class A approved systems to reduce fire spread potential of large vegetative roof areas."

§C4.2. Fire Protection for Roof Top Structures and Penetrations (p. 8)

"Pavers are often used as Class A or Non-Combustible separators. Care should be taken when installing pavers to avoid damaging the membrane. Some manufacturers require a separation material between the paver and the membrane."

§C4.3 Spread of Fire, Protection for Large Area Roofs (p. 8)

"This standard utilizes Area Dividers to reduce the potential for fire spread for large roof areas. Spread of flame for Class A fire is limited to 6 ft (1.8 m), if there is a 6 ft (1.8 m) break separating vegetative areas using Class A material or Non-Combustible Material the flame spread is not expected to ignite the nearby area. The dimensions chosen for large area roof-limitations are based on FLL requirements and FM Global recommendations (FM Global Loss Prevention Data Sheet 1-35—Green Roof Systems, they also coincide with the International Building Codes Area limitations for Assembly buildings.

FM Global has used ASTM E108 to test Vegetative Roofing Systems. Modifications of the test standards may be able to provide a meaningful test for selected conditions. However, with all the plant types that could be used in a roof design, the varying weather conditions that occur through the year, and the effects of seasons generate many variables that limit the potential to classify a roof construction. For this reason, Class A classified assemblies are limited to succulent based systems at this time. Refer to Green Roof Plants and Growing Media course manual, by Green Roofs for Healthy Cities, for definitions related to vegetative roof plant types.

The FLL believes that a vegetative "hard roof" can be considered to be equivalent to an ASTM E108 Class A Fire Classified roof assembly. The FLL defines a vegetative "hard roof" as those that are:

- irrigated;
- regularly maintained;
- have a substrate no less than 30 mm (1.18 in);
- made of vegetation that is grasses, succulents and/or perennials;
- have a substrate with at least 80% inorganic content by mass.

The agreed minimal substrate thickness varies between 30 mm (1.18 in) and 80 mm (3.15 in)."

Structural safety/Wind design

N/A.

Irrigation

There is not a dedicated section for irrigation requirements, but in §**5.0 Maintenance (p. 4)** it is recommended that irrigation should be provided: "provision(s) shall be made to provide access to water for permanent or temporary irrigation."

Moreover, in §**C5.0 Maintenance (p. 9)** it is specified: "One of the important ways for preventing fires is through the use of an Irrigation System. The need for irrigation will vary greatly due to climate and types of plants chosen."

Maintenance/Cost estimation

Maintenance is briefly addressed in §**5.0 Maintenance (p. 4)** , where it specifies:

"Maintenance shall be provided as needed to sustain the system by keeping vegetative roof plants healthy and to keep dry foliage to a minimum; such maintenance includes, but is not limited to irrigation, fertilization, weeding. Excess biomass such as overgrown vegetation, leafs and other dead and decaying material shall be removed at regular intervals not less than two times per year. Provision shall be made to provide access to water for permanent or temporary irrigation. The requirement for maintenance shall be conveyed by the designer to the building owner, and it shall be the building owner's responsibility to maintain the Vegetative Roof System. See Commentary C5.0."

The §**C.5.0 Maintenance (p. 9)** completes the maintenance requirements, by specifying: "The building owner needs to properly maintain a vegetative roof. One of the important ways of preventing fires is through the use of an Irrigation System.

The need for irrigation will vary greatly due to climate and types of plants chosen. Designers should be aware that plantings are to be specific for the roof being installed and that rooftops are at best hostile places for vegetation. Dead foliage should be removed and the moisture level of the Growing Media should be checked at regular intervals depending upon specific conditions on the vegetative roof. By regularly removing excess biomass that could become fuel for a fire on the rooftop, the risk of fire spreading beyond the 6 ft (1.8 m) Class A fire rated separation is minimized.

Best management practices for maintenance include regular weeding, fertilization, and removal of dead/dormant vegetation in accordance with the recommendations of the green roof provider. Specific directions for the proper maintenance of the vegetative cover should be furnished by the green roof provider."

Standard (5): ANSI/SPRI—American National Standard Institution/Single Ply Roofing Industry, 2016, ANSI/SPRI RP—14, Wind Design Standard for Vegetative Roofing System

The ANSI/SPRI RP—14 standard provides a method of designing wind uplift resistance of vegetative roofing systems utilizing adhered roofing membranes. It is intended to provide a minimum design and installation reference for those individuals who design, specify, and install vegetative roofing systems. It provides a method for defining the wind requirements of green roof systems and the type of ballast needed, based off of a set of measurable conditions (i.e., maximum wind speed, building height, parapet height, surrounding surface roughness, classification of buildings according to wind, snow and earthquake loads.).

Applicability

- Required wind uplift resistance and the minimum design/installation requirements for extensive green roofs and flat roofs;
- To be used in conjunction with the manufacturer's specifications and requirements for the green roofing system;
- Specifically, for green roofing systems in the United States, but can be used internationally.

Structure

The standard is divided into two main parts, each seven chapters in length: the first represents the standard itself, while the second part provides additional commentary to assist designers in applying the requirements of the preceding chapters:

Part 1: Standard:

1. Introduction of the standard object;
2. List of all the definitions used in the standard (i.e., ballast, growth media, etc.);
3. Various design considerations and system requirements, including physical dimensions of the building/roof (i.e., building height, roof slope, etc.), requirements for the components of the green roof system (i.e., membrane requirements, ballast requirements, etc.), and factors from the wind (i.e., erosion, wind-borne debris, etc.);
4. Explanation of green roof design options into three different types (i.e., a ballasted vegetative roofing system, a protective vegetative roofing

system, and a vegetative roofing system that uses a fully adhered membrane roofing), and further divides each type into three systems, describing in detail the corner area, perimeter, and field of the system;

5. Design provisions (i.e., large openings in the roof, perviousness of the decks, etc.);
6. Procedure to determine the green roof design, based on the characteristics from the previous chapters and the subsequent design table;
7. Brief description of maintenance requirements:
   * 14 tables and 5 figures defining the maximum allowable wind speed, depending on the building height, the system used, and the surface roughness
   ** Basic wind speed map of the United States, by ASCE, with the minimum design loads for buildings and other structures

Part 2: Commentary:

Explanatory and supplementary material to assist the designers and regulatory authorities in applying the standard (Part 2 follows the same chapter layout as Part 1).

Characteristics

The standard provides for a comprehensive method for determining the required wind uplift resistance applicable for a set of criteria nominated in the accompanying tables.

There are no references to vegetation type or height and is only applicable for extensive (not intensive) green roofs. Also, the basic wind speed map from ASCE is only applicable to the United States (although it could be applied in any other location, given the local wind speed).

About

Fire risk/Fire safety

N/A

Structural safety/Wind design

Characteristics and requirements for wind uplift resistance of the three different green roof systems are described in the sections in Table 1:

Table 1 Vegetative roofing systems implementable according to the standard.

System no [§Section, p.]	Description [§Subsections]
System 1 (§**4.2.1** **p.** 9)	"The installed membrane shall be ballasted with #4 ballast. See §**3.12.1**."
System 2 (§**4.2.2** **p.** 9)	"The installed membrane shall be ballasted as follows: §**4.2.2.1 Corner Area** The installed membrane in the corner area shall be ballasted with #2 ballast. See Section 3.12.2 and Figure 1. §**4.2.2.2 Perimeter** The installed membrane in the perimeter area shall be ballasted with #2 ballast. See Section 3.12.2 and Figure 1. §**4.2.2.3 Field** In the field of the roof, the installed membrane shall be ballasted with #4 ballast. See Section 3.12.1. For areas designated as wind debris areas, #2 ballast shall be the minimum size-weight ballast used."
System 3 (§**4.2.3** **p.** 9)	**"§ 4.2.3.1 Corner Area** In each corner area, the adhered roof system designed to withstand the uplift force in accordance with ASCE 7 or the local building code shall be installed in accordance with the provisions for the corner location with no loose stone, unprotected growth media, or unprotected modular vegetative roof trays placed on the membrane. See Figure 1 and Commentary C4.0. When a protective covering is required in the corner area, install minimum 22 psf (104 kg/m^2) pavers, or other material approved by the authority having jurisdiction. § **4.2.3.2 Perimeter** In the perimeter area, the adhered roof system designed to withstand the uplift force in accordance with ASCE 7 or the local building code shall be installed in accordance with the provisions for the perimeter location with no loose stone, growth media, or modular vegetative roof trays placed on the membrane. When a protective covering is required in a perimeter area, install minimum 22 psf (104 kg/m^2) pavers or other material approved by the authority having jurisdiction. §**4.2.3.3 Field** In the field of the roof, install #2 ballast. See §**3.12.2**."

Table 2 Ballasts applicable according to the standard.

#4 Ballast (§3.12.1 pp. 7-8)	**#2 Ballast (§3.12.2 p. 8)**
For vegetative roofs when vegetation nominally covers the visible surface of the growth media or provisions have been made to prevent wind erosion from the surface, #4 ballast can consist of any of the following used independently or in combinations: • Growth media spread at a minimum dry weight of 10 psf (49 kg/m^2) of inorganic material plus organic material. Interlocking contoured fit or strapped together trays containing growth media spread at minimum dry weight of 10 psf (49 kg/m^2) of inorganic material plus organic material; • Independently set modular preplanted or pregrown vegetative roof trays containing 18 psf (88 kg/m^2) dry weight inorganic material plus organic material. Vegetation coverage or erosion protection is not required when the #4 ballast below is used; • River bottom or coarse stone nominal 1-1/2 in (38 mm) of ballast gradation size #4, or alternatively, #3, #24, #2, or #1 as specified in ASTM D7655, Standard Classification for Size of Aggregate Used as Ballast for Membrane Roof Systems spread at a minimum weight of 10 psf (49 kg/m^2);	For vegetative roofs when vegetation nominally covers the visible surface of the growth media or provisions have been made to prevent wind erosion from the surface, #2 ballast can consist of any of the following used independently or in combinations: • Growth media spread at a minimum dry weight of 13 psf (64 kg/m^2) of inorganic material plus organic material; • Interlocking contoured fit or strapped together trays containing growth media spread at minimum dry weight of 13 psf (64 kg/m^2) of inorganic material plus organic material; • Independently set modular preplanted or pregrown vegetative roof trays containing 22 psf (104 kg/m^2) dry weight inorganic material plus organic material. Vegetation coverage or erosion protection is not required when the #2 ballast below is used; • River bottom or course stone nominal 2-1/2 in (64 mm) of ballast gradation size #2, or alternatively, #1 as specified in ASTM D7655 Standard Classification for Size of Aggregate Used as Ballast for Membrane Roof Systems spread at a

Continued

Table 2 Ballasts applicable according to the standard.—cont'd

#4 Ballast (§3.12.1 pp. 7-8)	#2 Ballast (§3.12.2 p. 8)
• Concrete pavers independently set (minimum 18 psf (88 kg/m^2)); • Interlocking, beveled, doweled, or contour-fit lightweight concrete pavers (minimum 10 psf (49 kg/m^2)).	minimum weight of 13 psf; (64 kg/m^2); • Concrete pavers independently set (minimum 22 psf (104 kg/m^2)); • Interlocking, beveled, doweled, or contour-fit lightweight concrete pavers (minimum 10 psf; (49 kg/m^2)).

The ballasts to which the three systems refer are described in the sections in Table 2.

These requirements of the three different systems are determined by the following factors:

(a) Basic wind speed: provided by the Standard, references ASCE 7—Basic Wind Speed Maps: (pp. 23-42), see Figure 1;

(b) Surface roughness category: there are three roughness categories provided for the surrounding ground surface, see Table 3;

(c) Building types classified according to wind, snow, and earthquake loads, see Table 3;

(d) Building height: there are 10 different height intervals, from less than 4.5 m (0–15 ft) to 41–46 m (135–150 ft), which affect the other factors. The building height is an input parameter in the tables of parapet heights, see Table 4;

(e) Parapet height of the roof: there are seven height intervals, defined through seven different design tables, from a 50–150 mm parapet height (2–6 inches) to parapets over 2 m in height (>72 inches), see Table 4;

(f) More design details are described through the several figures:
- The roof dimensions for the corner length and perimeter width, which is specified by the roof height, see Figure 2;
- Canopies and overhang eaves dimensions for impervious and pervious decks (Figures 2 and 3 of the standard, pp. 20-21);
- The parapet height design considerations (Figure 4, p.22 of the standard).

Attachment I
ASCE7-16 Figure 26.5-1A
Basic Wind Speeds for Risk Category I
Buildings and Other Structures

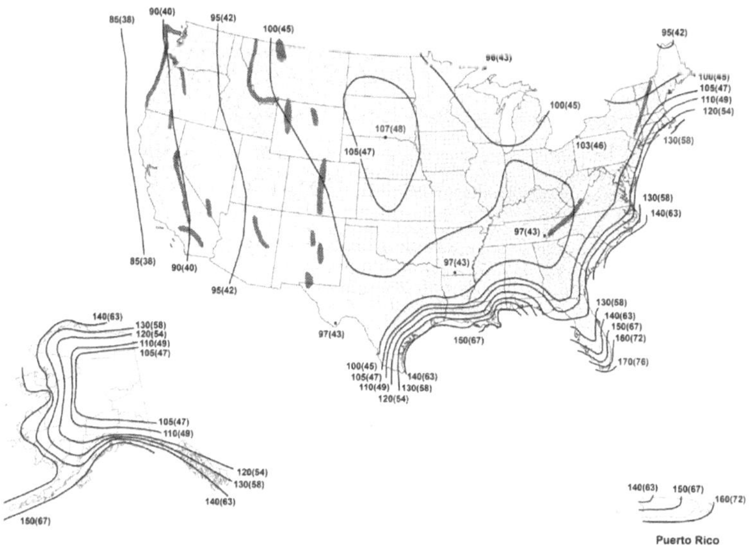

Location	V (mph)	V (m/s)
Guam	180	(80)
Virgin Islands	150	(67)
American Samoa	150	(67)
Hawaii	See Figure 26.5-2A	

ANSI/SPRI RP-14 2016
Wind Design Standard For Vegetative Roofing Systems

Approved
September 9, 2016

Notes: Dark shading indicates a Special Wind Region.

1. Values are nominal design 3-s gust wind speeds in miles per hour (m/s) at 33 ft (10 m) above ground for Exposure Category C.
2. Linear interpolation is permitted between contours. Point values are provided to aid with interpolation.
3. Islands, coastal areas, and land boundaries outside the last contour shall use the last wind speed contour.
4. Mountainous terrain, gorges, ocean promontories, and special wind regions shall be examined for unusual wind conditions.
5. Wind speeds correspond to approximately a 15% probability of exceedance in 50 years (Annual Exceedance Probability = 0.00333, MRI = 300 years).
6. Location-specific basic wind speeds shall be permitted to be determined using www.atcouncil.org/windspeed.

Figure 1 The Standards provides several maps to identify the basic wind speed of the project related to the risk categories of buildings (example p. 23).

Table 3 Surface Roughness/Exposure of the areas around the vegetative roof (pp. 3–4).

Surface Roughness/ Exposure category §27 **p. 3**	Description "A ground surface roughness within each 45-degree sector shall be determined for a distance upwind of the site as defined in Sections 2.7.1, 2.72, or 2.7.3 for the purpose of assigning an exposure category."
Surface Roughness/ Exposure B §**2.7.1 p. 3**	"Urban and suburban areas, wooded areas, or other terrain with numerous closely spaced obstructions having the size of single-family dwellings or larger. Exposure B: For buildings with a mean roof height of less than or equal to 30 ft (9.1 m), Exposure B shall apply where the ground surface roughness, as defined by Surface Roughness B, prevails in the upwind direction for a distance greater than 1,500 ft (457 m). For buildings with a mean roof height greater than 30 ft (9.1 m), Exposure B shall apply where Surface Roughness B prevails in the upwind direction for a distance greater than 2,600 ft (792 m) or 20 times the height of the building, whichever is greater."
Surface Roughness/ Exposure C §**2.7.2 p. 3**	"Open terrain with scattered obstructions having heights generally less than 30 ft (9.1 m). This category includes flat open country and grasslands. Exposure C shall apply for all cases where Exposures B or D do not apply. See Commentary C2.7."
Surface Roughness/ Exposure D §**2.7.3 p. 4**	"Flat, unobstructed areas and water surfaces. This category includes smooth mud flats, salt flats, and unbroken ice. Exposure D shall apply where the ground surface roughness, as defined by Surface Roughness D, prevails in the upwind direction for a distance greater than 5,000 ft (1,524 m) or 20 times the building height, whichever is greater. Exposure D shall also apply where the ground surface roughness immediately upwind of the site is B or C, and the site is within a distance of 600 ft (183 m) or 20 times the building height, whichever is greater, from an Exposure D condition as defined in the previous sentence. For a site located in the transition zone between exposure categories, the category resulting in the largest wind forces shall be used. See Section 5.3."

Table 4 Classification of buildings and other structure for wind, snow, and earthquake loads (Table 1, p. 5).

Nature of Occupancy	Category
Buildings and other structures that represent a low hazard to human life in the event of failure including, but not limited to, Agricultural facilities Certain temporary facilities Minor storage facilities	I
All buildings and other structures except those listed in Categories I, III, IV	II
Buildings and other structures that represent a substantial hazard to human life in the event of failure including, but not limited to, Buildings and other structures where more than 300 people congregate in one area Buildings and other structures with elementary school, secondary school, or day care facilities with capacity greater than 150 Buildings and other structures with a capacity greater than 500 for colleges or adult education facilities Health care facilities with a capacity of 50 or more resident patients but not having surgery or emergency treatment facilities Jails and detention facilities Power generating stations and other public utility facilities not included in Category IV Buildings and other structures containing sufficient quantities of toxic or explosive substances to be dangerous to the public if released including, but not limited to, **A.** Petrochemical facilities **B.** Fuel storage facilities **C.** Manufacturing or storage facilities for hazardous chemicals **D.** Manufacturing or storage facilities for explosives	III
Buildings and other structures designated as essential facilities including, but not limited to, Hospitals and other health care facilities having surgery or emergency treatment facilities Fire, rescue, and police stations and emergency vehicle garages Designated earthquake, hurricane, or other emergency shelters Communications centers and other facilities required for emergency response	IV

Continued

Table 4 Classification of buildings and other structure for wind, snow, and earthquake loads (Table 1, p. 5).—cont'd

Nature of Occupancy	Category
Power generating stations and other public utility facilities required in an emergency Ancillary structures (including, but not limited to, communications towers, fuel storage tanks, cooling towers, electrical substation structures, fire water storage tanks, or other structures housing or supporting water or other fire suppression material or equipment) required for operation of Category IV structures during an emergency Aviation control towers, air traffic control centers, and emergency aircraft hangers Water storage facilities and pump structures required to maintain water pressure for fire suppression Buildings and other structures having critical national defense functions	

[1]The definitions above are based on those of ANSI/ASCE 7-2010. Examples of building types are retained from previous version of ASCE 7 for clarification.

Table 5 Design Table A from 2 inch high to less than 6.0 inch high parapet – Maximum wind speed (MPH) (Table 2 p.12).

Roof height feet	System 1		System 2		System 3	
	Exposure C	Exposure B	Exposure C	Exposure B	Exposure C	Exposure B
0–15	110	115	125	125	140	140
15–30	110	115	120	125	140	140
30–45	100	110	110	125	140	140
45–60	No	No	105	125	130	140
60–75	No	No	100	120	130	130
75–90	No	No	No	No	No	No
90–105	No	No	No	No	No	No
105–120	No	No	No	No	No	No
120–135	No	No	No	No	No	No
135–150	No	No	No	No	No	No

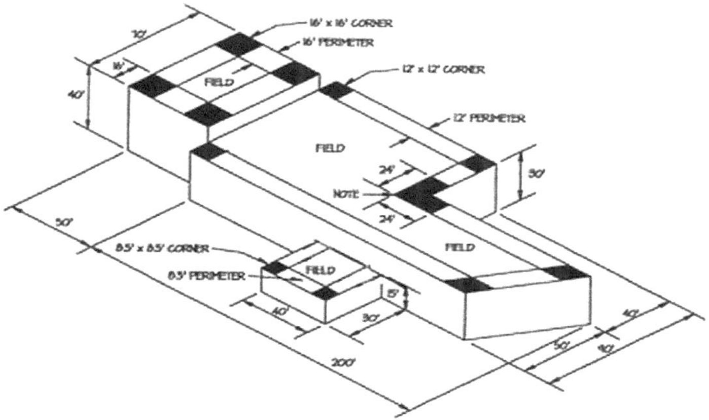

Note: Reentrant corners are larger than other corners.

	Low Roof	Main Roof	High roof
Roof Height	15 ft	30 ft	40 ft
40% of Building Height	6.0 ft	12 ft	16 ft
Corner Length	8.5 ft (a)	12 ft	16 ft
Perimeter Width	8.5 ft (a)	12 ft	16 ft

(a) 8.5 ft minimum controls

Metric Dimensions

	Low Roof	Main Roof	High Roof
Roof height	4.6 m	9.0 m	12 m
40% of building height	2.0 m	3.6 m	5 m
Corner length	2.6 m (a)	3.6 m	5 m
Perimeter width	2.6 m (a)	3.6 m	5 m

(a) 2.6 m minimum controls

Other Dimensions

Description	IP	Metric
	High Roof	
Corner	16 ft x 16 ft	5 m x 5 m
Perimeter	16 ft	5 m
Width	70 ft	21.3 m
Height	40 ft	12 m
	Main Roof	
Corner	12 ft x 12 ft	3.6 m x 3.6 m
Perimeter	12 ft	3.6 m
Height	30 ft	9 m
Re-entrant Corner	24 ft x 24 ft	7.3 m x 7.3 m
Off set	40 ft	12 m
Width	90 ft	27.4 m
Length	200 ft	61 m
	Low Roof	
Corner	8.5 ft x 8.5 ft	2.6 m
Perimeter	8.5 ft	2.6 m
Width	30 ft	9 m
Height	15 ft	4.6 m

Figure 2 Roof Areas System 2 and 3 (pp. 18–19 of the standard)

Irrigation

N/A

Maintenance/Cost estimation

N/A

Standard (6): CSA Group—Canadian Standards Association, 2015, A123.24-15 Standard test method for wind resistance of modular vegetative roof assembly

This standard provides requirements for wind resistance testing of modular vegetated roof assemblies, based on research and tests conducted by the National Research Council Canada (NRCC). This standard was prepared and formally approved by the Technical Committee on Bituminous Roofing Materials, under the jurisdiction of the Strategic Steering Committee on Construction and Civil Infrastructure.

Applicability

- Specimen setup, instrumentation and apparatus, testing procedures, and termination requirements for extensive green roofs, with a modular vegetated assembly, are supplied;
- The test method is for modular vegetated systems that are grown off-site and placed on the roof to achieve coverage, but not for mechanically attached membrane roofing systems. A modular vegetated system, roofing system, and vegetated roof assembly are defined as follows:
  - Modular vegetated system: a system of pregrown or precultivated vegetation (modules, blankets, or mats), growth media, root barrier, pavers, and a drainage system, with at least 80% vegetation coverage;
  - Roofing system: a system consisting of a deck and roofing or water-proofing membrane, vapor barriers or retarders, insulation, cover boards, etc.;
  - Vegetated roof assembly (VRA): the combination of a roofing system and modular vegetated system assembled together.

Structure

The standard is divided into 10 paragraphs and 2 annexes:

1. Description of the scope of the standard;

2. List of references;
3. Definition of modular vegetated system, roofing system, and vegetated roof assembly (VRA);
4. Purpose of the test;
5. Precautions for to consider when conducting the test;
6. Requirements for a wind uplift resistance test for a VRA (i.e., test apparatus, test specimen, instrumentation);
7. Requirements for a wind flow resistance test for a VRA (i.e., test apparatus, test specimen, instrumentation);
8. Testing procedures and data inherent the test wind speed used in the testing shall be determined by building code requirements or by the manufacturer/client;
9. Description of the termination of the tests;
10. Description of the test report (i.e., the index of the report).

Annex:

A VRA wind design process with the three steps of the test;

B Sample test report (an example of the final report of the test).

Characteristics

The test procedure in this standard is clear and comprehensively described.

About
Fire risk/Fire safety

N/A.

Structural safety/Wind design

As described in §**4.1 Significance of the test (p. 8)**: "The purpose of the test is to evaluate the wind resistance of vegetated roof assembly under laboratory-generated dynamic pressure and air flow conditions. The result of this test is useful in determining the suitability of a vegetated roof assembly on a particular building in a particular geographical location in relation with the building code."

For the wind uplift resistance and wind flow resistance tests, the standard specifies:

- Test apparatus
- Test specimen
- Instrumentation
- Test procedures
- Termination of the test

	Wind uplift resistance test	Wind flow resistance test
Test specimen	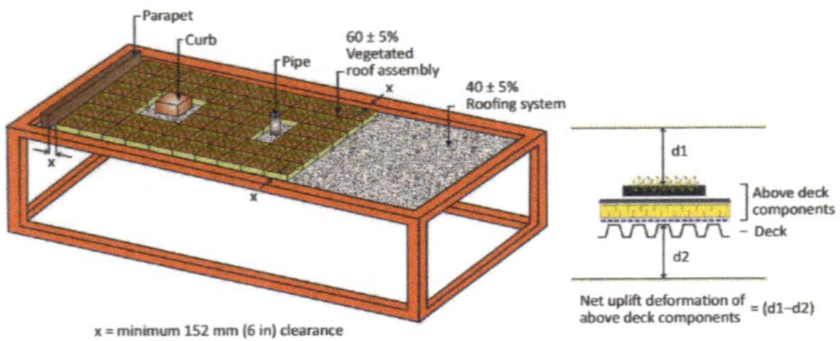 Figure 1 **§6.2 Test specimen** "The test specimen of vegetated roof assembly consists of a combination of roofing system and modular vegetated system. They respectively cover 40 ± 5% and 60 ± 5% of the test specimen area or as specified by the authority having jurisdiction. The segment in which the modular vegetated system is installed shall have a minimum of 152 mm (6 in) spacing (vegetation-free zone) around its perimeter and shall include at least two penetration details (curb and stack pipe) and a parapet. A minimum of two vegetated modules/blankets or mats shall be installed surrounding the details."	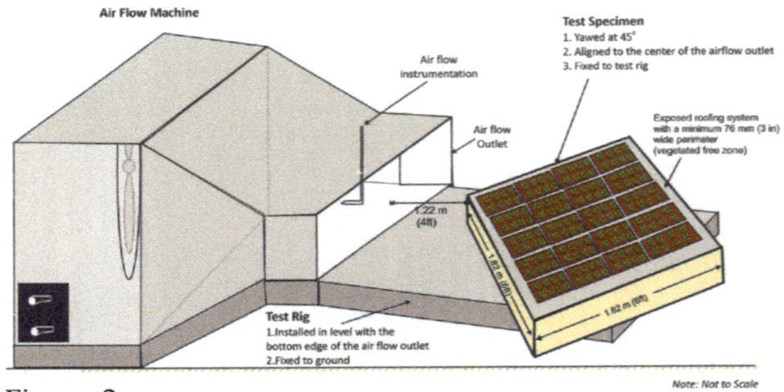 Figure 2 **§7.2 Test specimen** **§7.2.1** "The roofing system shall be installed with a minimum slope of 2% on a mock-up size of 1.82m (l) 1.82m (w) 0.30m (h) (6 ft 6 ft 12 in) and cured according to the manufacturer or client requirements. A grid pattern of 25.4 mm 25.4 mm (1 in 1 in) shall be marked on the membrane top surface. These visible marks will be used during the testing to identify any sliding or overturning movements (see Figure 2). Note: Care should be taken to minimize any air intrusion into the roofing system through edges during the wind flow testing by applying air sealing methodologies." **§7.2.2** "A vegetation-free perimeter zone 76 mm (3 in) wide shall be maintained, and the modular vegetated system shall be installed over the balance of the roofing system area according to the manufacturer or client's design and instructions (see Figure 2)."

		§7.2.3 "The test rig shall be installed in level with the bottom edge of air flow outlet (see Figure 2) and fixed to the ground to provide sufficient rigidity during wind flow testing. The completed vegetated roof assembly test specimen should be yawed at 45° to the air flow machine simulating corner wind effects and shall be positioned at 1.22m (4 ft) from the air flow outlet aligned to the centre (see Figure 2). The test specimen shall be fixed to a test rig to provide sufficient rigidity during wind flow testing."
Instrumentation	**§6.3 Instrumentation** **§6.3.1** "A pressure-measuring apparatus, capable of measuring the test pressure differential with a ± 0.05% accuracy of full-scale pressure or ±50 Pa (±1.0 psf), whichever is smaller, shall be used. A pressure sensor shall be installed at each of the following three locations: (a) on the inside of the chamber, to indicate the reference pressure; (b) on the top of the vegetation, to measure the simulated pressure; (c) on the top of the membrane, to measure the pressure on the roofing system." **§6.3.2** "A deflection-measuring apparatus capable of measuring the induced deflection with a ±0.05% accuracy of full-scale deflection or 0.01 mm (±0.004 in), whichever is smaller, shall be used. A minimum of two sensors shall be installed; one inside	**§7.3 Instrumentation** "Air flow instrumentation capable of continuously measuring and recording velocity accurately to ± 1 kmph (0.6 mph) shall be used."

Continued

	Wind uplift resistance test	Wind flow resistance test
	the test chamber at the centre of the vegetation to quantify the assembly uplift movement (d1), and the other at the centre of the structural span between joists (L) to measure the structural deck deformation (d2). The net uplift deformation of the above-deck components including vegetation shall be recorded as (d1−d2)."	
Test procedure	§**8.1** "The wind uplift rating of vegetated roof assembly shall be evaluated in accordance with the procedures described in CSA A123.21."	§**8.2.1** "Calculation of test wind speed (V): The test wind speed used in the testing shall be determined according to building code requirements for a specific building, or it shall be a speed specified by the manufacturer or client, whichever is greater." §**8.2.2** "With the test specimen in position, the test shall start with test wind speed, V, and shall be continuous when moving from one speed level to another with increments of 10 kmph (6.2 mph) up to failure or up to a minimum of 1.5V, or as specified by the manufacturer or client. Note: Calibration of the air flow measurement sensors to the data recording system can be verified by running the air flow machine at a sample wind speed of 48 kmph (30 mph)."
Termination of the test	§**9.1.1** "The test shall be carried out until the failure of the test specimen as described in Clause 9 of CSA A123.21 or when the net uplift deformation of the above-deck components exceeds 2L/240, where L is the deck structural span."	§**9.2.1** "The test shall continue until the wind speed specified by the manufacturer or client has been sustained, or until such time as the test specimen exhibits signs of failure. Failure shall be considered to have occurred if any of the following is noticed: (a) there is overturning

	§**9.1.2** "When uplift deformation is used as the failure criteria, the corresponding test pressure at L/240 shall be referred to as the "sustained pressure" and the corresponding test pressure at 2L/240 shall be referred as the "failed pressure". Alternatively, the manufacturer or client may report the wind uplift resistance obtained from CSA A123.21 as the sustained pressure."	of any components within the test assembly; (b) there is sliding or displacement of any components of the test assembly that is greater than 25.4 mm (1 in)." §**9.2.2** "The speed at which failure occurs shall be referred as "failed speed", and the incremented speed level before failed speed shall be referred as the "sustained speed"."

Irrigation

N/A.

Maintenance/Cost estimation

N/A.

Standard (7): ASTM International–American Society for Testing and Materials International, 2015, ASTM E2400 Standard Guide for Selection, Installation, and Maintenance of Plants for Green Roof Systems

This standard guide provides characteristics for green roof systems and how they can perform in conditions, such as exposure to strong or variable wind forces and/or sunlight/radiant heat. Due to extreme environmental factors, especially at height, the selection of the appropriate plant material is vital to the success of a green roof system.

Applicability

- A general guide for the selection, installation, and maintenance of plants for extensive and intensive green roof systems.

Structure

The standard is divided into nine parts:

1. Scope of the standard guide;
2. Referenced documents (i.e., E631 Terminology of Building Construction, E2114 Terminology for Sustainability Relative to the Performance of Building);
3. Referenced documents for the terminology (i.e., E631, E2114);
4. Summary of the guide;
5 Explanation of how to use this standard to benefit green roof systems;
6. Characteristics of plants and lists criteria to determine appropriate selection of plant species for intensive and extensive green roofs (i.e., design intent, esthetics, climate, microclimate, roof slope and orientation, plant characteristics, etc.);
7. Information for propagating and installing plant material in green roof systems (i.e., pre-cultivation, direct planting on roof);
8. Information about maintenance (i.e., irrigation, periodic fertilization);
9. Summary of the keywords.

Characteristics

This standard represents a valid introduction for plant selection of green roofs. The language is simple, understandable, especially for professionals who do not deal with plants often. Also, §**6.2 Criteria for Selecting Species** is very valuable.

This guide is only meant to provide general guidance and specifies that consultation with a professional horticulturist, green roof consultant, or similar professional that is familiar with green roof technology and plants is necessary.

About

Fire risk/Fire safety

N/A.

Structural safety/Wind design

N/A.

Irrigation

Irrigation is explained in §**8.1 Irrigation (pp. 3–4)**, which is a subsection of §**8 Maintenance**. The following summarizes this section:

- General information: "Container-grown plants and plugs should be watered immediately following planting and then watered frequently for the first few weeks unless ample rainfall occurs. Then gradually wean them off of irrigation. After establishment during the first year, watering may not be necessary depending on the local environment and the plant species chosen. The amount of water required depends on the plant species. In general, extensive roofs should not require irrigation. However, it may be advisable to have an irrigation system in place just in case the site experiences an extended drought or other unusual weather patterns."
- Passive versus active irrigation: "[...] Passive systems store rainwater in the drainage layer, which moves back up into the media layer via capillary action. Examples of passive systems include fiber mats placed under the growing media that act like a sponge; cups formed into the drainage mat; and contours in the bottom of a modular unit.

Active irrigation involves pumping water onto the plants either by sub-irrigation or, more commonly, overhead irrigation. [...] Watering may be performed manually, controlled by time clocks, or computer programs that sense media moisture. In all cases, water must be monitored.

An irrigation system should be customized to each specific green roof system based on the intended use of the roof, choice of plant materials, media depth, environmental conditions, and budget. A professional irrigation specialist should perform irrigation system design."

- "The need for long-term irrigation is dictated by climatic conditions at the site, especially the amount and distribution of rainfall and temperature extremes. These conditions can eliminate the possibility of using certain species."

Maintenance/Cost estimation

§8 **Maintenance (pp. 3–4)** provides general information about irrigation (previously explained) and plant maintenance. §**8.2 Plant Maintenance (p. 4)** specifies that green roofs, especially intensive green roof systems, require regular maintenance. The subsections of §**8.2** are summarized below:

- §**8.2.1**: "Periodic fertilization may be necessary to maintain lush growth. Species such as *sedum* require very little fertilizer and often do better when neglected."
- §**8.2.2**: "Herbicides, insecticides, fungicides, and fertilizers should be avoided. If they are necessary, they should be used sparingly. These chemicals could potentially hasten degradation of the roof membrane. During the first two growing seasons, undesired plants should be removed so that the desired vegetation can obtain full coverage. Volunteer woody species such as poplar, should be removed from extensive roofs before the roots compromise the roof materials. Weeds can be controlled by utilizing shallow media layers and by foregoing irrigation. Many weeds cannot survive in shallow medium depths and thus have trouble competing with drought tolerant species such as *sedum* or *sempervivum*."
    - §**8.2.2.1**: "If limited funds are available for long-term maintenance, then drought tolerant succulents such as *sedum* should be grown in a relatively shallow media. Otherwise, weeds may become a serious problem. If the maintenance budget is not a concern, then the choices of plant material and media depths are widely expanded."

CHAPTER 5

Data sheets from insurance company

Data sheets (1): FM Global, 2011, Property Loss Prevention Data Sheets 1-35—Green Roof Systems

This document is developed to reduce exposure to natural hazards, fire, and excessive loads for green roofs. It represents a complementary instruction to an exhaustive standard for green roof systems, so it is created to be used in conjunction with other standards. It often references other standards, especially FM Global Data Sheet 1-28 Wind Design.

Applicability

- These data sheets collect useful information/recommendations for preventing and reducing green roof exposure to natural hazards, fire, and excessive loads (including wind).

Structure

The standard is divided into six main chapters and five appendixes and includes a list of figures:

1. Scope of the standard;
2. Loss prevention recommendations: classification of the green roof systems and list all the recommendations for wind, wind speed restrictions, building height restrictions/loads, supporting structure/fire exposure, minimum foreseeable loss fire walls, fire breaks/roof drainage, roof slope/green roof components/maintenance;
3. Support for recommendations: provides better explanations for the recommendations listed in Chapter 2;
4. References: a complete list of the references from the previous chapters, divided into two sections (FM Global and other);

Green Roofs, Facades, and Vegetative Systems
ISBN 978-0-12-817694-8
https://doi.org/10.1016/B978-0-12-817694-8.00005-4

Appendix

1. Glossary of Terms;
2. Document Revision History;
3. Supplementary Information;
4. Bibliography;
5. Sample Green Roof Project Photos.

Data Sheets 1-35 provide detailed instructions for what wind risks, fire, and excessive loads. These instructions are further given extensive explanations in §**3.0 Support for recommendation (pp. 13–17)**.

The section on operations and maintenance does not go into complete detail in some aspects.

About

Fire risk/Fire safety

The fire topic is treated in §**2.2.12 Fire Exposure (p. 9)** and is summarized in the following table:

Section	Instructions	References
§**2.2.12.1 Exterior Fire Exposure (p. 9)**	Due to the many variables (including plant type, plant condition, depth of growth media, combustibility of roofing assembly materials, and installation details) and the lack of sufficient experience and test data, classification of exterior fire exposure cannot be made with certainty at the present time.	
§**2.2.12.2 Interior Fire Exposure (p. 9)**	Evaluate green roof systems for interior fire exposure (as regards to a Class I or Class II rating) in the same manner as for conventional roofing systems on metal deck. For green roof systems on concrete deck, assume the assembly is non-combustible with regard to interior fire exposure.	

Section	Instructions	References
§2.2.13 **MFL Fire Walls (p. 9)**	Provide areas free of vegetation and growth media adjacent to MFL walls in conformance with the minimum setbacks as defined in Data Sheet 1-22, Maximum Foreseeable Loss. Extend vegetation-free border zones not less than 50 ft (15 m) on each side of an MFL wall and cover with stone ballast, concrete paver blocks (refer to Data Sheet 1-29, Roof Deck Securement and Above-Deck Roof Components), or a gravel surfaced roof cover as noted in Data Sheet 1-22. Refer to Data Sheet 1-22 for additional provisions regarding roof surfacing and parapets at MFL walls.	• Data Sheet 1-22, Maximum Foreseeable Loss • Data Sheet 1-29, Roof Deck Securement and Above-Deck Roof Components
§2.2.14 **Non-Vegetated Border Zones, Fire Breaks, and Parapet Walls (p. 9)**	Provide stone ballast or concrete paver blocks to cover all border zones that are designated to be free of vegetation and growth media. Refer to Data Sheet 1-29, Roof Deck Securement and Above-Deck Roof Components, for minimum recommendations regarding ballasted systems; note that the provision for increased minimum dimension (8.5 ft [2.6 m]) of perimeter and corner areas as described in §**2.2.7.4** of Data Sheet 1-29 need not apply to green roof systems unless the roofing membrane relies on ballast to resist wind uplift pressures.	• Data Sheet 1-29, Roof Deck Securement and Above-Deck Roof Components

Continued

Section	Instructions	References
§**2.2.14.1 Perimeter and Border Zones (p. 9)**	Ensure border zones (as defined by perimeter and corner zones in Data Sheet 1-28, Wind Design) are free of vegetation and growth media.	• Data Sheet 1-28, Wind Design
§**2.2.14.2 Rooftop Structures, Penetrations, and Joints** (subsections: §**2.2.14.2.1,** §**2.2.14.2.2,** and §**2.2.14.2.3) (p. 9)**	§**2.2.14.2.1** Provide a minimum 1.5 ft (0.5 m) wide continuous border zone (free of vegetation and growth media) surrounding all rooftop equipment, penetrations (e.g., ducts, drains, pipe, conduit), skylights, solar panels, antenna supports, expansion joints, roof area dividers, and interior parapet walls (unless part of an MFL Fire wall; refer to §**2.2.13**, MFL Fire Walls**,** for additional requirements). Consider wider vegetation-free zones in cases where HVAC intake or exhaust could be expected to affect or be affected by plant growth; for instance, where taller plant growth might restrict intake, or high velocity exhaust emissions could cause plant damage. §**2.2.14.2.2** Provide a minimum 3 ft (0.9 m) wide continuous border zone (free of vegetation and growth media) around rooftop structures, including but not limited to mechanical and machine rooms, penthouses, and adjacent façade walls. §**2.2.14.2.3** Provide 3 ft (0.9 m) wide continuous border zone strips (free of	

Section	Instructions	References
	vegetation and growth media) to partition the roof area into sections not exceeding 15,625 ft^2 (1450 m^2), with each section not exceeding 125 ft (39 m) in length. Incorporate the border zones into expansion joints or roof area dividers wherever possible.	

In §**3.10 Non-Vegetated Border Zones (p. 15)**, several recommendations that clarify the role of border zones are listed as follows:

(1) To provide maintenance access because green roof vegetation is not intended to support foot traffic;

(2) To provide additional resistance to high wind uplift pressures;

(3) As a means of reducing scour of growth media;

(4) To reduce the potential generation of wind-borne debris at roof perimeters and corners;

(5) To provide a fire break at rooftop equipment, penetrations, and structures.

Structural safety/Wind design

The wind topic, which is treated in §**2.2 Construction and Location (pp. 5–13)** and explanations are provided in §**3.0 Support for Recommendations (pp. 13–17)**, is summarized in the following table:

FM Global Property Loss Prevention Data Sheets 1-35—Green Roof Systems

Section	Instructions	§3.0 Support for recommendations
§2.2.1 Building Height Restrictions (p. 5)	• For buildings over 150 ft (46 m): use concrete pavers in non-vegetated border zones; not use roof gravel or stone ballast. • For buildings not exceeding 150 ft (46 m): stone ballast or concrete pavers may be used in non-vegetated border zones.	

Continued

FM Global Property Loss Prevention Data Sheets 1-35—Green Roof Systems

Section	Instructions	§3.0 Support for recommendations
§2.2.2 Wind Speed Restriction (p. 6)	"Install green roof systems only in geographical locations where the basic wind speed (3-second gust) is less than 100 mph (45 m/s). This applies to all building heights."	**§3.1 Wind Speed Restrictions** "The geographic locations of green roof systems are limited based on wind speed (3-second gust, as determined by Data Sheet 1-28, Wind Design) in order to eliminate or reduce the likelihood that growth media and/or roof gravel will become wind borne debris. The limitation is also intended to reduce the effects of wind-induced scour of growth media and the resulting loss of ballast of above-membrane components, as well as potential damage to the vegetation."
§2.2.3.2 Wind Uplift, Roof Ballast, and Safety Factor (p. 6)	See following sections.	
§2.2.3.2.1 (p. 6)	"Use a mechanically attached or fully adhered roof membrane system based on appropriate wind uplift design pressures and wind rating pressures as determined from Data Sheet 1-28, *Wind Design*."	

FM Global Property Loss Prevention Data Sheets 1-35—Green Roof Systems

Section	Instructions	§3.0 Support for recommendations
§2.2.3.2.2 (p. 6)	"Use growth media (engineered soil) as ballast against wind uplift for the roofing membrane and other waterproofing elements only where a uniform depth of 8 in (200 mm) or more is provided; when this is the case, use a minimum safety factor of 1.7 for wind uplift calculations based on appropriate wind uplift design pressures as determined from Data Sheet 1-28, *Wind Design*. Calculate the safety factor based on a dry condition (no water present in the growth media, retention mat, drainage panel, etc.) and without the presence of any vegetation load."	**§3.2 Wind** "The growth media on a green roof system can be prone to scour from wind and water action and therefore may not be a reliable source of uniform ballast for waterproofing components when provided in shallow depths. Refer to Data Sheet 1-29, *Roof Deck Securement and Above-Deck Roof Components,* for stone ballast requirements in non-vegetated border zones. Note that stone ballast is not to be used on roofs greater than 150 ft (46 m) in height."
§2.2.3.2.3 (p. 6)	"Use a minimum of 3 in. (76 mm) of stone ballast unless greater depths are recommended either in Data Sheet 1-29, *Deck Securement and Above-Deck Roof Components,* or by the manufacturer or installer. Use clean, smooth, well-rounded stone ballast that conforms to the gradation requirements of Standard Size No. 3 Course Aggregate per ASTM D 448 (nominal 1 to 2 in. [25 to 50 mm] in diameter)."	

Continued

FM Global Property Loss Prevention Data Sheets 1-35—Green Roof Systems

Section	Instructions	§3.0 Support for recommendations
§2.2.3.2.4 (p. 6)	"Growth media may be used as a secondary ballast material; that is, it may be used to ballast the loose-laid roofing components above the waterproofing membrane (i.e., drainage panel, retention mat, root barrier, and insulation board), but not the membrane itself. Use wind uplift design pressures (as determined from Data Sheet 1-28, *Wind Design*), a dry condition (no water present in growth media, retention mat, drainage panel, etc.), no vegetation, and a minimum safety factor of 0.85 for wind uplift calculations where growth media is used to ballast the loose-laid above-membrane components. See §**2.2.16.2**, Growth Media, for additional minimum depth requirements of growth media."	**§3.3 Wind-Borne Debris (p. 13)** "Growth media typically contains aggregate materials, including crushed porous rock (e.g., pumice, expanded shale) or crushed clay brick, that can create a potential source of wind-borne debris similar to roof gravel. Although vegetation will provide a certain amount of wind shelter and the plant roots will help anchor the surrounding growth media, it can take several growing seasons for the vegetation to become sufficiently established to protect the growth media from wind action."
§2.2.3.2.5 (p. 6)	"Pre-cultivated vegetated mats (rather than direct planting of plugs or cuttings) are sometimes used for green roofs. Where vegetated mats are installed, anchor them until the mat's root growth has achieved sufficient attachment into the growth media to adequately resist wind action (at least one full growing sea- son). Ensure vegetated mats are properly	

FM Global Property Loss Prevention Data Sheets 1-35—Green Roof Systems

Section	Instructions	§3.0 Support for recommendations
	anchored or ballasted against wind forces based on a safety factor of 1,0. If pre-cultivated vegetated mats are to be used as ballast for the roofing membrane or other waterproofing elements - refer to section **§2.2.3.2.2** for safety factor requirements."	
§2.2.3.3 Wind-Borne Debris (p. 6)	"Do not use woody vegetation on a green roof when the wind uplift pressure at the roof elevation is equal to or greater than the uplift pressure at an elevation of 15 ft (4.5 m) or less for ground roughness B and a basic wind speed (3-sec gust) of 110 mph (49 m/s)."	
		§3.10 Non-vegetated border zones (p. 15) "Non-vegetated border zones are recommended for several reasons: **(1)** to provide maintenance access because green roof vegetation is not intended to support foot traffic; **(2)** to provide additional resistance to high wind uplift pressures;

Continued

FM Global Property Loss Prevention Data Sheets 1-35—Green Roof Systems

Section	Instructions	§3.0 Support for recommendations
		(3) as a means of reducing scour of growth media; **(4)** to reduce the potential generation of wind-borne debris at roof perimeters and corners; **(5)** to provide a fire break at rooftop equipment, penetrations, and structures."

Irrigation

The irrigation topic is briefly explained in §**2.3.1 Irrigation (p. 12)** (part of §**2.3 Operation and Maintenance**, which is one of the less extensive sections). It is specified how to design the irrigation for extensive and intensive green roofs:

- §**2.3.1.1**: For green roof systems designed to be self-sustaining without regularly scheduled irrigation (i.e., most extensive systems), provide rooftop hose bibs to allow for irrigation during initial stages of plant propagation and during occasional drought conditions.
- §**2.3.1.2**: For green roof systems that require regular irrigation (i.e., intensive and simple intensive systems), ensure that a permanent irrigation system is provided.

The hose bibs, installed in an extensive green roofs, provide an access to water in case of fire, although this function is not quoted.

Maintenance/Cost estimation

Just a few points are made for maintenance in §**2.3 Operation and Maintenance (pp. 12–13)**, which are summarized below:

- §**2.3.2 Fertilization, Pesticides, and Plant Care**:
  - §**2.3.2.1**: Fertilization will most likely be necessary during the initial establishment of the vegetation over the first and second growing

seasons; after that time, occasional fertilization, perhaps once or twice annually, might be necessary. Consult the roofing manufacturer to determine which chemical fertilizers and pesticides, if any, are acceptable and will not damage the waterproofing assembly or void the warranty.

  - §**2.3.2.2**: It includes inspections and selective weeding as part of the regular maintenance work and provides these at least twice annually.

- §**2.3.3 Leak Test and Inspection**
- §**2.3.4 Warranty** (for vegetation and waterproofing)

CHAPTER 6

Vegetative Roofs Guidelines

Guidelines (1): FLL—Forschungsgesellschaft Landschaftsentwicklung Landschaftsbau e.V. The Landscaping and Landscape Development Research Society e.V., 2018, Guidelines for the Planning, Construction and Maintenance of Green Roofing

The FLL guidelines for green roofing are the most extensive code of practice in the world, exclusively dedicated to green roofs. The first version of the document, the *Principles of Green Roofing* (1982), was reworked in 1992 as the *Guidelines for Planning, Construction and Maintenance of Green Roofing* by the Landscape Development and Landscape Research Society in Germany. It has been updated since this time, and the most recent international edition is the 2018 (English version).

The "modern green roof technology," with its multilayer system, was first developed in Germany and then exported in the other European Countries and all around the world. In Germany, the *Guidelines for Planning, Construction and Maintenance of Green* has served as the accepted regulations for green roof design for buildings. Thus, the FLL Green Roofs Guidelines are internationally known as the most extensive design guide and has served as a benchmark for other countries developing their own standards. It is the primary point of reference for all guidelines produced globally.

Applicability

- Green roof systems, including intensive, simple intensive, and extensive greening for terraces, halls, underground car parks, etc.;
- General green roof requirements and specific requirements for each green roof layer;
- Necessary technical information to develop professional agreements between clients, planners, and contractors.

Green Roofs, Facades, and Vegetative Systems
ISBN 978-0-12-817694-8
https://doi.org/10.1016/B978-0-12-817694-8.00006-6

Structure

The standard is divided into seventeen chapters plus three appendixes:

1-2 Scope of the guidelines and the normative references
3 Definitions
4 Legal framework conditions
5 Types of green roofs and forms of vegetation
6 Function and effects
7 Requirements for construction and building materials
8 Structural requirements
9 Requirements for the construction of vegetation areas
10 Drainage layer
11 Filter layer
12 Vegetation stratum
13 Requirements for seeds, plants, and vegetation
14 Planting and seeding
15 maintenance services
16 Acceptance, claims for defects
17 Testing

Appendix A: Informative orientation values for load assumptions and water storage

Appendix B: Investigative methods for vegetation substrate and drainage bulk materials for green roofs

Appendix C: Procedure for investigating the root penetration resistance of membranes and coatings for green roofs

Characteristics

These guidelines are recognized by its high level of completeness, including very detailed technical information for green roof systems as a whole and for each individual component and layer. Furthermore, it serves as a guarantee of the quality and success of the system for the client and all involved in the design, construction, maintenance, development, and production/manufacturing.

About

Fire risk/Fire safety

The fire topic is developed in §**8.9 Fire prevention measures (pp. 44–45)**: the guidelines define green roofs as resistant to flying sparks and radiant heat if they are intensive and, in case of extensive, if they respond to a number of characteristics listed below.

Rules for the construction and use phases are added.

"With regard to fire prevention measures, the demands made on roofs in state building regulations mean that they must be sufficiently resistant to fire exposure from the outside due to flying sparks and radiant heat ('hard roofing'), otherwise a number of constraints, in particular greater distances to adjacent buildings, must be adhered to.

In DIN 4102 Fire behavior of building materials and building components—Part 4: Synopsis and application of classified building materials, components, and special components, green roofs are classified as 'roofs are resistant to flying sparks and radiant heat' if they have the following characteristics:

(1) Intensive green roofs are considered as roofs that are resistant to flying sparks and radiant heat.

(2) Extensive green roofs are resistant to flying sparks and radiant heat if they have the following characteristics:

- Mineral-rich vegetation layer with maximum 20% (by weight) of organic constituents;
- Vegetation stratum with a layer thickness $\geq$30 mm;
- Façade walls, firewalls, or walls that are permissible instead of firewalls must finish, at intervals of not more than 40 m, at least 0.3 m above the roof, relative to the top of the vegetation layer. If these walls do not finish above the roof, a 0.3 m high upstand of noncombustible building materials or a 1 m wide strip of solid slabs or coarse gravel is sufficient;
- A spacer strip of solid slabs or coarse gravel $\geq$0.5 m wide shall be formed around openings in the roof surface (domed rooflight, skylights) or to rising walls with windows, if its balustrade is $\leq$ 0.8 m above the vegetation stratum;
- In the case of aligned, gable-ended buildings, a horizontal strip of at least 1 m wide strip must remain ungreened along the eaves and provided with surface protection made of noncombustible building materials." (DIN 4102-4 2016, para. 11.4.7, **p. 197**).

In addition, during construction and use, general rules of fire prevention must be observed, such as

- Construction phase
  - orderly construction site conditions;
  - no storage or rapid disposal of packaging material;
  - quantity and time limited storage of building materials according to construction progress;

- decentralized storage of building materials such as waterproofing membranes and geotextiles;
- careful handling of open flames and other heat sources (e.g., gas burners and hot air welding equipment for laying membranes, grinders).

• Use
  - maintenance of the defined condition "hard roofing" through appropriate care and maintenance (e.g., watering, removing dried plant parts, keeping spacer strips free of vegetation);
  - make any necessary changes in case of a change of use from the defined state "hard roofing";
  - no control of unwanted vegetation by means of heat appliances (e.g., gas burners, hot air devices, infrared devices);
  - careful handling of open fire and other heat sources (for example, garden grill, campfire, radiant heater).

The §**8.6.2.2 Marginal strips (p. 42)**: the marginal strips in gravel are introduced as elements useful for fire prevention. In this case, "[...] if the marginal strip is to function as a safety strip for preventive fire protection, it must be designed in accordance with the requirements of DIN 4102-4 (§ 8.9)."

Structural safety/Wind design

The structural safety and wind topic are developed crosswise to all the guidelines, referring to different layers/components and different aspects of designing green roofs.

Among the sections under §**7 Requirements for construction and building materials (pp. 28–34)**, the topics of loads and falling materials are introduced (see §**7.6 Load assumptions (p. 31)** and §**7.7 Protection against falling (p. 31)**).

Then, important requirements for the stability and the safety of green roofs are discussed among the sections under §**8 Structural requirements (pp. 35–53)**. In particular, §**8.6 Transition points (pp. 41–43)** mention transition points of the roof according to location (like rising structural elements, doors, barrier-free transitions, roof penetrations, roof edge) and provide instructions and reference to other standards. §**8.6.2.2 Marginal strips (p. 42)** is particularly important to prevent wind suction (and fire).

§**8.10 Protection against material displacement (pp. 46–50)** widely discusses the three forms of material displacement: surface erosion, slipping of layers at layer boundary, material displacement when the bulk angle of repose is exceeded.

In §**8.11 Edging (p. 51)** and §**8.12 Accessible surfaces (pp. 51–51)**, the stability of edging and surface material is an explicit requirement.

§**8.13 Furnishings (pp. 52–53)** underlines the need to solve case by case the design and the structural issue.

In §**9 Requirements for construction of vegetation areas (pp. 55–63)**, §**9.3 Water retention (pp. 57–60)** states that "the maximum water capacity is to be considered structurally in addition to the load of the layer structure at maximum water capacity" **(p. 59)**.

§**14.3 Securing the stability of woody plants (p. 89)** explains how to achieve stability of big shrubs and trees through bracing and anchoring.

Appendix A lists all the loads of green roof materials (in kg/m^2 × 1 cm thickness) and the water storage capacity (expressed in l/m^2 × 1 cm thickness).

Following there are all the interesting excerptions, most of which are mentioned above.

In §**5.3.3 Structure-dependent factors (p. 24)** under the §**5.3 Determination of site conditions for vegetation**, among several factor to be considered in the design phase, the following are simply listed with no further explanation: wind flow conditions; additional water load from adjoining structural elements; the gradient or pitched roof surfaces and length; design load and resulting depth of structured layers; additional technical installation like air conditioning units, antenna, solar panels; puddle forming of roofs.

In §**7.6 Load assumptions (p. 31)**, some cautions about loads are given:

"The static conditions, i.e., the load assumptions, are a limiting selection criterion for determining the type and construction of a green roof.

The structure with all its layers at maximum water capacity, including the areal load of the vegetation, is to be classified as part of the permanent effect. The load from the water stored during the ponding process is also to be considered. The point loads of large shrubs, trees, and structural elements, e.g., pergolas, water basins, edging elements, are to be determined separately and taken into account accordingly (Appendix A).

The sufficient compressive strength of the thermal insulation in connection with the roof waterproofing must be taken into account, in particular in the production of the green roof and the arrangement of point loads.

If the layers of the green roof structure are to serve as a load for the wind suction protection of the underlying roof structure, see §8.8."

In §**7.7 Protection against falling (p. 32)**, suitable safety measures are mentioned. Only tested products and systems are recommended:

"Health and Safety accident prevention measures are to be taken into account during the planning and tendering stages for the building. This applies specially to fall prevention during execution, maintenance and servicing of buildings, and the fall-through prevention for building elements (Skylights). Corresponding requirements include the construction site ordinance (BaustellV), the DIN 4426, the DGUV regulation 38 with BGV/GUV-V C22 DA, the DGUV information 201-056, as well as the accident prevention regulation of the 'Unfallverhütungsvorschrift der Sozialversicherung für Landwirtschaft, Forsten und Gartenbau (Gartenbau-Berufsgenossenschaft) VSG 4.2'.

Suitable safety measures can be, e.g., barriers, anchorage for rope safety systems, permanent lifts with fall protection, and entry options. Only tested products or systems are to be used. The responsibility for the safety measures lies with the client and his assigned planner or safety coordinator and safety officer.

During implementation work, fall protection in the form of balustrades on the roof or a façade scaffold is usually necessary. Particular attention should be paid to the coordination of the various trades, so that, for example, temporary safety scaffoldings are dismantled only after completion of the green roof.

For care and maintenance work, personal safety equipment against falling (PSAgA) is usually sufficient and appropriate anchor devices have to be provided by the client. For installation on roof waterproofing that has already been completed, it is advisable to attach the anchorage points without penetrating the roof skin, e.g., by attachment to rising structural elements or safety systems held by ballast."

In §**8.6.2.2 Marginal strips (p. 42)**, "The marginal strips in gravel are introduced as required elements against wind suction, besides inspection areas to waterproofing membrane.

Moreover, the marginal strips may be considered functional to fire prevention (according to DIN 4102-4).

The marginal strip on green roofs is usually, mostly, vegetation-free, e.g., gravel, aggregate, or slabs, and forms a space between the transition and the vegetation area. The marginal strip serves the visual inspection of the waterproofing, the edge of the roof, and rising structural elements.

The marginal strip can be designed as a safety strip against wind suction, whereby the width, the area load, and the wind blow protection are to be specified during planning (§8.8).

If the marginal strip is to function as a safety strip for preventive fire protection, it must be designed in accordance with the requirements of DIN 4102-4 (§8.9).

If bulk materials are used, they should consist of particle sizes 16/22 to 16/32 unless otherwise specified. They may have particles $\leq$8 mm of at most 5% by mass, <0.063 mm at most 3% by mass. In the case of bulk materials consisting mainly of rounded particles (e.g., river gravel), proportions of cracked particles do not represent a defect."

In §**8.6.3.1 Transitions to rising structural elements (p. 42)** under § **8.6.3 Implementation**, marginal strips in gravel are defined as safety strips (and slash protection) in case of transitions with façades and other rising structural elements:

"At transitions with façades and rising structural elements, the marginal strip serves as a safety strip and splash protection. It must be passable for inspection, maintenance, and care work. The spacing to the vegetation prevents plants from growing in behind the waterproofing connection and from being impaired in their development by water running off the façade or dripping. If the strip has to be used for cleaning the façade, it has to be made correspondingly wider.

The edge of the roof waterproofing/root barrier membranes shall be fixed above the surface of marginal strips, vegetation areas, and accessible coverings in accordance with the requirements specified in Section 8.6.2.1. [...]

If vegetation areas are created at a distance from the façade, different construction options are available:

- continuous drainage layer and, if necessary, filter layer under the vegetation stratum and marginal strip;
- vegetation stratum and margins next to each other without edging or separated by embedding, e.g., perforated plates;
- demarcation and separate drainage of vegetation stratum and margins;
- Installation of drainage channels instead of the marginal strip."

The § **8.8 Protection against negative wind pressure (pp. 43–44)** discusses the topic of negative wind pressure. Reference to German standards are quoted as well as important safety rules: "Due to the effects of the wind, structures are subjected to pressure, suction and friction forces, the intensity of which depends on the wind direction and the shape and height of the building. The wind loads occurring in the roof area can cause damage to the roof structure during construction as well as in a finished condition. Protection measures for the roof waterproofing and the layers belonging to the roof structure against lifting due to wind loads, must be pre-defined during the roof planning phase.

According to DIN EN 1991-1-4, roof areas are divided into different areas according to their different levels of wind stress. In particular, the ridge, corner, and edge areas are exposed to higher loads. They are to be secured by appropriate measures.

If the roof waterproofing and root barrier are loosely laid, they must be secured against lifting wind forces by the layer structure of the green roof. In edge and corner areas, e.g., bulk gravel material and in higher load areas, slabs should be installed. During construction, temporary safeguard measures are required.

In general, when constructing green roofs, the aim is to keep the load and the height of the layer structure as low as possible, it may be necessary to secure the particularly vulnerable edge and corner areas by increasing the layer thicknesses or by using heavier materials. The load of the layer structure in a dry state is decisive here. If necessary, the securing of the edge and corner areas must be carried out in combination with, e.g., gravel, grass stones, or slabs.

The load required for securing can be calculated in accordance with DIN EN 1991-1-4 and in conjunction with the corresponding partial safety factors according to DIN EN 1990. Information on the safeguarding of roof waterproofing against lifting by wind forces can also be found in the ZVDH/HDB 'Fachregel für Abdichtungen—Flachdachrichtlinie'.

These requirements, based solely on the weight of gravel or concrete slabs, do not take into account that additional positive factors are effective for green roofs:

- the roughness of the vegetation;
- the load of residual moisture in the layer structure;
- the load of the vegetation stock;
- the fixation of a large area through extensive root penetration of the entire layer structure in comparison to the single-particle composition of a gravel layer;

- the wind permeability of the vegetation stratum, which leads to a pressure equalization between the top and bottom of the layer and thus to a load reduction.

For the design according to DIN EN 1991-1-4, the aerodynamic coefficient for the external pressure cpe 10 can be used.

An extended section is dedicated to prevent the materials displacement in green roofs.

The §**8.10 Protection against material displacement (pp. 46–50)** explains the problems related to "surface erosion, slipping and exceeding the angle of repose."

8.10.1 Types (p. 46): "Fundamentally, three forms of material displacement are to be differentiated between green roofs, for which different safeguarding measures are partly necessary. These include

- surface erosion (see §8.10.1.1);
- slipping of layers at a layer boundary (see §8.10.1.2);
- material displacement when the bulk angle of repose is exceeded (see §8.10.1.3).

Depending on the exposure, measures against surface erosion may be required for roofs with a pitch of as little as 0°. Measures against slipping may be required as of 10°, depending on the design or construction. Dependent on the substrate, measures may be required against exceeding the angle of repose."

§**8.10.1.1 Surface erosion (p. 46)**: "Surface erosion can occur on both sloped and flat roofs. In particular, wind and water attack during the production of green roofs is a threat to the vegetation stratum, the sprouts, and the seed. The period of danger is shortened by the introduction of vegetation in a favorable season for the vegetation development.

Until the layer structure is fixed by the infiltration of the roots of the vegetation, there is a risk to the vegetation stratum, plants, and seeds by exposure to water and wind.

To counteract damage,

- special requirements for the layer structure (choice of material, chippings of hard rock as a mulching layer);
- temporary effective measures (erosion control fabric, soil stabilizer, keep damp);
- as well as special measures for extreme locations (wet sowing, vegetation mats) may be necessary, some of which may be combined (see §**8.10.2** and §**8.10.3**).

The risk of erosion of substrates by wind is not to be judged by the wind speed, but by the material-related rate of drift.

In the case of pitched green roofs in a sun-exposed position, in addition to erosion protection affected by the vegetation, irrigation may be necessary at times along with a nutrient supply."

§8.10.1.2 Slipping of layers at a layer boundary (p. 47): "The risk of slippage of the vegetation structure or of individual layers on pitched roofs exists at each layer boundary and depends on the friction between the materials. Depending on the design or construction, safeguarding measures may be necessary above a pitch of 10°. Optimization options include reducing the number of layers or using composite elements (e.g., drainage elements with bonded protective geotextiles).

To prevent slipping, the following measures can be implemented:

- increasing the friction by matching the materials to each other;
- constructive support at the eaves;
- constructive suspension at the ridge;
- constructive measure in the roof area (e.g., shear ties)."

§8.10.1.3 Material displacement when the bulk angle of repose is exceeded (p. 47): "Displacement of bulk material occurs when the natural angle of repose of the material is exceeded during installation or in the long term. On sloping roofs, this is problematic if the maximum angle of repose is smaller than the roof pitch."

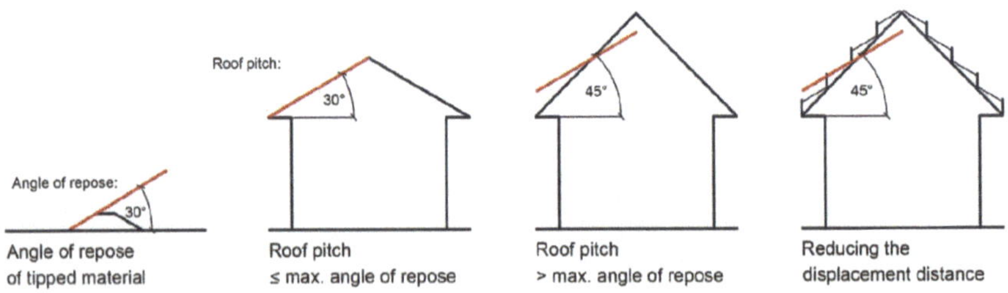

Figure 1: Surface formation of the substrate as a function of angle of repose, roof pitch, and when installing shear ties (Fig.2 **p. 47** of the guidelines).

"The resulting possible surface slope depends on the substrate and cannot be influenced by measures against slipping. Structurally stable substrates can be installed on surface slopes of up to 30° without additional measures (see 2).

To avoid the displacement of bulk material above the maximum angle of repose, the following measures are possible:

- Increasing the shear strength in the bulk material (e.g., plant-compatible adhesives, fibers, or vegetation);
- Reduction of displacement distances (e.g., grid, chamber, or honeycomb systems);
- Bundling bulk materials in (e.g., wire-mesh boxes, textile bags, cassettes);
- Surface coverage (e.g., through fabric-reinforced vegetation mats).

Alternatively, bulk-material-free constructions can be used."

§**8.10.2 Material properties (p. 48)**: "In order to protect against surface erosion, in particular against negative wind pressure on flat roofs and on sloping roofs, structurally stable growing media with higher individual particle weights (particle size/density) can also be used in the dry state. Further structural stability (against surface erosion and exceeding the angle of repose) can be achieved by intermeshing particle shapes in the fine and medium gravel range as well as small amounts of elutriates and organic matter.

Areal materials are to be matched in their friction properties.

Constructive measures of whatever form, e.g., statically effective connected mesh mats, ties, grid profiles, studded panels, fabrics, and their fasteners, must be durable and weather resistant.

They must not hinder the vegetation development. Degradable materials (such as wood) should be excluded as a permanent safeguard.

The tensile strength of geotextiles or geotextile composites and, where applicable, their fasteners shall be designed for the static loads. In addition, the snow load must also be taken into account.

Anti-erosion fabrics must remain functional until the completion care has been carried out.

Plant species and vegetation and cultivation forms should be suitable for the site as well as enable a quick and permanent area coverage."

§**8.10.3 Manufacture (pp. 49–50)**: "Irrespective of the pitch, measures can be necessary to protect against water and wind erosion on all roof surfaces until the root system has developed. Precultivated vegetation mats, dense vegetation, erosion protection fabrics, wet seed plants, and sprouting seeds in combination with mulch and glue can be used here. If wind erosion is to be expected, it is also possible to keep the vegetation substrate constantly wet during the completion care.

In the case of (wind) erosion-prone substrates, greening with vegetation mats or turf stones should be carried out on the roof edges and on particularly vulnerable subareas.

For roofs with a slope of more than 3°, additional measures can be necessary, according to DIN 18531-3 and the ZVDH/HDB 'Fachregel für Abdichtungen—Flachdachrichtlinie', such as bonding of the roof waterproofing, installation of support structures, and mechanical fastening to prevent slipping of the layer package.

In order to reduce the sliding layers on roofs with a roof pitch of more than than 10°, penetration proof waterproofing is preferable to a separate waterproofing and root barrier membrane. If in exceptional cases this cannot be guaranteed, the positional security of the root barrier membrane must be ensured and proven.

Above a roof pitch of 10°, safeguarding measures against slipping are required. These measures may, in the simplest embodiment, consist of drainage elements, if in the form of rolled material and having the tensile strength, being placed over the ridge or attached to it. If the construction is inherently rigid, the drainage elements can be supported on the eaves. It. must be determined that the ridge and/or eaves can bear the resulting loads and the required attachment is sufficiently strong. Loosely laid geotextiles above drainage elements should be avoided above this pitch.

For roofs with an inclination of 20° and above, a reduction in the number of sliding layers is necessary and construction methods with loosely laid geotextiles should be avoided, i.e., these roofs can be constructed in single-layer format or with drainage elements without geotextiles.

Furthermore, at this roof pitch and above, erosion protection must be ensured directly and independently of exposure.

For roofs with a pitch of more than 30°, there is a risk of slipping as well as a displacement of bulk material. For measures against slipping at this pitch and higher, a separate static calculation is required. The load transfer to constructive safety elements, eaves, and other roof installations must be proven. Manufacturers of constructive safeguarding elements should provide appropriate structural calculations or calculated values. The safeguarding measures must not add additional tension to the waterproofing/root barrier.

Measures against the displacement of bulk material according to paragraph 8.10.1.3 may be used for slopes of 30° and above. If none of these safeguarding measures is implemented, the client must be informed of the possibility of displacement.

Green roofs on roofs with a slope of more than 45° should be avoided due to the increasing risk of slipping and displacement of bulk material."

The §**8.13 Furnishings (pp. 52—53)** sets all the requirements for furnishing applied to green roofs:

§**8.13.1 Types (p. 52)**: "Furnishing elements are, for example, as follows:

- trellises;
- pergolas;
- lamps;
- pools.

The arrangement and design of furnishing elements are object-specific, detailed solutions that are to be solved structurally, statically, and physically on a case-by-case basis."

§**8.13.2 Requirements (p. 52)**: "Furnishing elements must be stable, load-distributing, and, if necessary, anchored. The tensionless positioning on the ground is of particular importance. The occurring point and/or area loads as well as the wind loads are to be taken into account."

§**8.13.3 Implementation (p. 53)**: "The attachment of furnishing elements can be done by

- load-distributing anchoring on the roof with constructive pre-installation or
- flat or grid-like foundations (see §**8.4**).

The constructive pre-installation is possible with foundations lifted above of the waterproofing level. In addition to the static requirements, the specifications for the formation of roof penetrations must be observed (see §**8.6**).

Constructive retrofitting should be the exception. Roof waterproofing/root barrier membranes and thermal barrier layers with vapor barriers must not be broken. The formation of flat or grid-like foundations requires the application of a sliding and protective layer over the underlying layers of the roof structure. The dimensioning depends on the type of furnishing and the effect.

The subsequent manufacture of roof penetrations for the anchoring of components, which are to be provided for the use of the roof space, should be avoided as far as possible. If constructive foundations for anchorage were not planned, it is necessary to check whether construction and furnishing elements, e.g., pergolas, trellis, lamps and benches, can be anchored to overlying and load-distributing foundation plates or grid foundations."

The §**14.3 Securing the stability of woody plants (pp. 89–90)** provides instructions to fix big shrubs and trees:

§**14.3.1 Requirements**: "The stability of larger trees and shrubs can be achieved through

- bracing and
- anchoring.

Measures for securing by bracing and anchoring serve the temporary stability assurance of the trees and shrubs. It is assumed that the specified minimum layer thicknesses are not undercut and a sufficiently rootable substrate volume is present.

During the working life of anchors and bracing, the trees and shrubs must be regularly checked for constrictions, pressure, and chafing damage."

§**14.3.2 Tension (p. 89)**: "Wire and rope bracing should be provided with a tightening mechanism. They can be attached

- with threaded anchors directly on the structure above the waterproofing. The connection is to be in a screwable form;
- to building elements with appropriate structural and static properties, e.g., edging, walls, large-sized floor panels;
- to building elements embedded in the green roof structure, e.g., ground panels, mesh mats. The permissible load-bearing capacity of the elements and/or the thermal insulation and roof waterproofing must not be exceeded. The bracing to the elements should not exceed an angle of 60°."

§**14.3.3 Anchoring to support frames (p. 90)**: "Square or triangular frames are suitable for anchoring trees. They are to be made of tubular steel with corrosion-resistant surface protection. The individual supports must be provided with load-distributing support plates."

Irrigation

Irrigation recommendations are very briefly discussed in §**7.9 Irrigation (p. 33)**:

"Irrigation shall be provided by at least one water connection on the roof, which is protected against frost by technical installation or maintenance.

The dimensioning and number of supply lines and connections as well as the required water pressure for irrigation are dependent on the local situation, the structural conditions, size and layout of the object, and the intended vegetation form and are to be considered in the planning. The specifications according to DIN EN 1717 are to be observed. See also FLLBewässerungsrichtlinien—Richtlinien für die Planung, Installation und Instandhaltung von Bewässerungsanlagen in Vegetationsflächen."

Moreover, the topic is explained in the §**9.4 Water storage and additional irrigation (p. 61)**:

"The vegetation development on green roofs is significantly influenced and limited by the water supply. If, for structural reasons and economic constraints, the load assumptions and structural thicknesses are to be kept as low as possible, the water storage space and thus the water supply are considerably reduced.

The water storage can take place in different layers and in different ways. Depending on the construction and arrangement in the individual layers, the following types of water storage are to be distinguished:

- in the vegetation stratum by using water-storing materials in the growing media and, if necessary, additionally in water storage panels or mats;
- in the vegetation stratum and additionally in the drainage layer by using particle-graded, open-pored bulk materials or prefabricated, draining, water-storage panels;
- in the vegetation stratum and additionally in the drainage layer by areal accumulation in bulk materials or under special drainage panels or partial accumulation in drainage profiled panels;
- and possibly additionally in water-storing protective layers and protective plies.

Simultaneous water storage in the vegetation stratum and drainage materials can be used in all types of green roof. The entire layer structure is intensively rootable space and is available for water storage.

In the case of intensive greening, the combination of water storage in the vegetation stratum with water accumulation in the drainage layer is an expedient form of water storage.

In simple intensive roof areas with a thin-layered construction, water accumulation in the drainage layer only makes sense if additional irrigation takes place in low-precipitation periods to avoid drought damage to the vegetation.

Ponding in extensive green roof areas brings with it disadvantages from a plant physiological point of view (§**7.3**)."

§**9.4.2 Additional irrigation (pp. 61–62)**: "The use of free and freely available rainwater forms the basis of the water supply to green roofs in order to minimize the need for additional irrigation and to introduce the rainfall directly back into the natural water cycle. Additional irrigation is regularly required for intensive greening if necessary and may be required for extensive greening during the completion care. For more steeply inclined roofs, the possibility of additional irrigation should be foreseen in order to avoid plant failures in extreme periods of drought thus avoiding the onset of erosion."

Maintenance/Cost estimation

Maintenance recommendations are clearly developed in §**15 Maintenance services for the development and maintenance of vegetation (development and maintenance care), maintenance (pp. 92–95)**. The section is divided into four subsections: §**15.1 General (p. 92)** that discusses the important topic of maintenance contracts, §**15.2 Intensive greening (pp. 92–92)**, §**Extensive greening (p. 94)**, and §**15.4 Maintenance (p. 95)**.

Following are the exceptions of the sections dealing with maintenance:

§**15 Maintenance services for the development and maintenance of vegetation (development and maintenance care), maintenance (pp. 92–95)**

§**15.1 General (p. 92)**: "The maintenance services for the development and maintenance of vegetation (development and maintenance care) are defined in DIN 18919, described with individual measures for ground-based greening and transferred to intensive greening accordingly.

For extensive green areas and partially also for simple intensive green areas, maintenance objectives and individual measures are coordinated with the greening method, the vegetation form, the biodiversity or the created habitat as well as the state of development and the direction of development for the object. The necessary waterproofing inspections remain unaffected.

The range of greening objectives lies between aesthetic ornamental gardens on one side and ecologically optimized green roofs on the other. Through the ongoing natural development, these pictures lose their clarity over the years. This can also be a desired effect. Otherwise, it will be necessary to intervene with suitable maintenance measures. The following describes all possible measures that come into question. The choice, type, and extent are to be specified by a professional and are object-specific.

Maintenance contracts, with the definition of a long-term conception under specialist supervision by the planning and monitoring landscape architects and/or implementing contractors that extend beyond the defects limitation period, are strongly recommended for both intensive and extensive green roofs.

In the contractual terms and conditions, the individual tasks should be clearly defined. These may be described by the type, extent, timing, and duration according to area (m^2) (e.g., vegetation area, edging/safety margin, surface covering), number or length (e.g., irrigation and drainage facilities). They may also be agreed upon as required.

During all works on roofs above a height of 3 m, protection against falling is required and the corresponding regulations governing the accident prevention have to be respected (see 7.7).

Where façade cleaning is required, the vegetation and layer structure will need to be protected against harmful substances before work starts."

The table below summarizes the contents of the subsections of the §**15.2 Intensive greening (pp. 92–92)**, §**15.3 Extensive greening (p. 94)**, and §**15.4 Maintenance (p. 95)**.

Table 1: List of maintenance activities.

	Intensive greening	Extensive greening	Servicing
List of tasks	It may be necessary to carry out the following tasks: • Loosening and cleaning plant beds; • Removing unwanted vegetation, especially germinating woody plants; • Removing refuse; • Fertilizing; • Watering; • Plant protection; • Pruning; • Mulching; • Removing fallen leaves; • Protective measures against frost and snow;	The following tasks are necessary: • Supply of nutrients; • Irrigation steeply pitched roofs, sun-exposed roofs; • Removing woody seedling and other unwanted growth; • Pruning for thinning; • Seeding in the case of larger barren areas with seeds or shoots of sedum species; • Replanting for larger areas without sufficient coverage;	The following technical facilities are to be maintained. In particular, the following must be ensured: • The functioning of the roof drains and the technical equipment for drainage and/or irrigation housed in inspection shafts. • The removal of dirt and deposits in inspection shafts, on pop-up sprinklers and in roof drains, as well as drainage channels. • The stability of balustrades,

Continued

Table 1: List of maintenance activities.

	Intensive greening	Extensive greening	Servicing
	• Tightening anchors and bracing; • Removing unnecessary anchoring and bracing; • Checking irrigation systems; checking the drainage system; • Keeping safety margins and paving free from dead leaves and overgrowing vegetation. Extra attention necessary for lawn and meadow-like areas: • Mowing; • Removing grass clippings; • Removing refuse; • Fertilizing; • Watering; • Plant protection; • Scarifying; • Aerating; • Sanding; • Repeat seeding; • Checking irrigation systems; • Checking the drainage systems; • Keeping safety margins and paving free from dead leaves and overgrowing vegetation.	• Replacing substrate that has eroded; • Refilling of substrate in cases of erosion; • Plant protection; • Keeping the technical equipment free of vegetation; • Keeping edging and safety margins and paving free from dead leaves and overgrowing vegetation. More tips: • The natural development dynamics of the vegetation on extensive green roofs begins after completion services (completion care). This can be influenced to a minimal extent through maintenance interventions, e.g., pruning measures or removal of individual plants. Migrating, high-growing, repressive plant species, e.g., some legumes (Fabaceae) should be removed at an early stage.	surface fastenings, and other structural elements. • The continued functionality of antislip devices on pitched roofs. • Functionally impairing deposits in gravel strips, at entrances and closures as well as in gravel beds at technical facilities are to be removed at intervals of several years.

Table 1: List of maintenance activities.

	Intensive greening	Extensive greening	Servicing
		The maintenance services for the development of vegetation (development care) for extensive green roofs are carried out after the completion services (completion care) for a limited period, until a projective soil coverage of about 90% is achieved. This can extend over two years, depending on the greening procedure and the state of development. In particular, with growing media for single-layer constructions and more steeply sloping roofs, the nutrient supply must be ensured for this period. • Gravel edges and safety strips as well as paving	

Continued

Table 1: List of maintenance activities.

	Intensive greening	Extensive greening	Servicing
		and slab joints can self-fertilize. Low-growing plants, e.g., moss, sedum, and cushion-growing herbs and perennials, should be tolerated. Unwanted spontaneous growth, especially in the case of vegetation-free fire protection strips, must be eliminated by regular maintenance, which must be agreed in terms of type, extent, and frequency. • In extensive greening, the supply of nutrients can usually be limited to the temporary period of maintenance services for development. It is recommended to apply a NPK long-term fertilizer on a basis of 5 g N/m^2 and year. • In a nutrient-poor structure, e.g., single-layer	

Table 1: List of maintenance activities.

	Intensive greening	Extensive greening	Servicing
		and thin-layer constructions, it may be further necessary to undertake a maintenance fertilization at intervals of several years to maintain the desired vegetation and flowering aspect.	
Frequency	8–10 times/year Lawn and meadow-like areas: 2–12 times/year	2–4 times/year	Multiple-year intervals

Guidelines (2): CIBSE—Chartered Institution of Building Services Engineers, 2007, Knowledge Series 11 Green Roofs

The document provides information about several issues on green roofs, such as legislation, retrofitting, cost, maintenance, and esthetics. The distinctive characteristic of the text is it is written in a Question/Answer format, as shown in the image:

Applicability

- This publication is aimed at design professionals that are designing green roofs;
- It provides an overview of the advantages of green roof systems and their implication on the project.

Structure

The standard is divided into nine parts:

1. Introduction and contents of the publication
2. Overview of green roofs: layering and technical information, benefits, climate mitigation, etc. (In this section it specifies that there are no

British Standards regarding green roofs and that the best option is to utilize the FLL standard)

3 Design considerations: planning, structural consideration, SuDS-Sustainable urban Drainage Systems potential, biodiversity, fire prevention, roof orientation and pitch, offset of cooling plant, compatibility with solar thermal or PV panels, reuse of rainwater from a green roof, and cost implication

4 Constructing green roofs: materials management, time management, access for construction, and health and safety

5 Maintaining green roofs

6 Retrofitting green roofs, in terms of structural implications, cost implications, suitable buildings, and health and safety

7 Costs and financial aspects

8 Legislations, regulations, and planning policies about green roofs

9 References

Characteristics

The document provides a wide range of information, including technical data (i.e., the thickness and weight of common materials for each layer), information about performance (i.e., reducing storm-water runoff as part of a Sustainable urban Drainage Systems-SuDS strategy), and a section dedicated to costs. The document also provides useful information to increase the general knowledge about green roofs and for making the decision of which system to use with more consideration.

About

Fire risk/Fire safety

§3 **Design considerations for green roofs (pp. 15–16)** takes the fire risk into consideration:

> Are green roofs a fire risk? There is a potential fire risk during prolonged hot weather when the vegetation is dry. Although there are at the present no mandatory fire standards for green roofs, it is recommended to follow the FLL standard, which recommends the use of pebble vegetation barriers of fire-breaks. The FLL recommendations further state that extensive roofs are unlikely to be ignited by sparks provided that:

- Substrate depth exceeds 30 mm and contains no more than 20% organic content by weight;
- Pebble or concrete paving fire breaks are provided in the vegetation every 40 m with a minimum width of 1 m or a 300 mm high fire wall;
- Vegetation barriers are provided to all roof penetrations and in front of walls with openings;

- Vegetation barriers are kept clear of encroaching plants by routine maintenance.

In §5 **Maintaining green roofs (p. 23)**, vegetation breaks are mentioned as an important safety element to prevent the spread of fire. They must be maintained at their original width and cleared of any encroaching plants.

Structural safety/Wind design

N/A.

Irrigation

§3 **Design considerations for green roofs (pp. 15–16)**:

Watering is recommended considering the type of green roofs and the stage of planting. Watering may be needed for:

- Intensive green roofs;
- Semi-intensive and extensive green roofs while establishing;
- Every type while establishing in case of south faces and drier parts of the United Kingdom.

> Will a green roof need to be irrigated? This depends on the type of roof but generally access to a water point should be provided in the design for most green roof installations. Intensive green roofs will need considerable watering throughout their life. Simple intensive and extensive green roofs may require some watering while establishing, but after that should not need watering unless there is a prolonged period (usually defined as six weeks) without rain. To reduce the need for irrigation, a green roof should be installed in the spring or autumn. Some green roof suppliers recommend watering sedum blankets during the establishment stage on south-facing roofs with steep pitches wherever they are installed; they also recommend watering during the establishment stage for any green roof if installed in the drier parts of the UK (i.e., where the average yearly rainfall is 650 mm or less). A biodiverse roof is designed to require no watering. It is important to realize that water supplies are under increasing stress in many parts of the UK and there is a considerable carbon load to water use (0.6 kg CO^2 per m^3). Serious consideration should therefore be given to collecting rainwater on site if specifying any type of green roof that will require watering.

§5 **Maintaining green roof** underlines the need of water access during the establishment stage:

> Access to water point is essential during the establishment stage of most green roof. This is of further importance to larger green roofs and those that have been designed with flora, which demands more regular irrigation. After an extensive sedum green roof is established, it should not require watering unless there is a drought. As a period of drought could coincide with a hosepipe ban, it is recommended that rainwater harvesting should be

incorporated if specifying a green roof in water stressed areas. During a drought the requirement for watering will result in increased maintenance.

Maintenance/Cost estimation

The following table collects the questions regarding maintenance in §5 **Maintaining green roofs (pp. 23–25)**:

Question	Answer
Who is responsible for maintenance?	Reputable contractors will include, or at least suggest and offer, an agreed period of regular maintenance. The period is usually between 12 and 18 months immediately following fitting and after the first full growing season.
What are the relevant health and safety implication during maintenance?	Fall protection must be provided.
When should maintenance be carried out?	Yearly basis, normally each spring.
What to inspect for?	Any roof clutter or blockage of drains is of primary concern, as is unwanted foreign growth, which could be detrimental to the plant species originally established, biodiversity aims and even the roof structure itself.
Are weeds problem?	If weeds are not allowed, they can be cleared. After 3–4 weeks, the cutting will be established. Better if they are cut in spring or autumns.
Is irrigation necessary on regular basis?	Access to water point is essential during the establishment stage of most green roof. This is of further importance to larger green roofs and those that have been designed with flora, which demands more regular irrigation. After an extensive sedum green roof is established, it should not require watering unless there is a drought. As a period of drought could coincide with a hosepipe ban, it is recommended that rainwater harvesting should be incorporated if specifying a green roof in water stressed areas. During a drought the requirement for watering will result in increased maintenance.

Question	Answer
Should fertilizers be used on a green roof?	Not fertilizing is recognized as the preferred option as, although specialist slow-release fertilizers can help a green roof to establish by promoting plant growth, the variety of species will be reduced and their use will result in increased nutrient levels in storm-water runoff which will negatively affect local water quality.
Is plant growth or vegetation breaks problem?	Vegetation breaks have an important safety function to prevent the spread of fire.
Is it possible to damage the roof structure during maintenance?	Tools must be carefully chosen so as not interfere or damage anything below the substrate.
Can the drainage points and flood the roof?	All drainage points must be checked every year and cleared out if necessary to ensure optimum performance.

§3 Design considerations for green roofs "Is access for maintenance required?" (p. 16) underlines the importance of the access to green roof for maintenance.

The topic of green roof costs is deepened in the following sections:

§3 Design considerations for green roofs "What is the cost of a green roof?" (p. 19): "A green roof will cost more than a conventional roof. Costs depend on the type of green roof specified and it is important to remember that the £/m^2 cost of a small green roof will be greater than that of a large one. However, specifying green roofs can reduce the amount of final attenuation capacity required in a SuDS system, and the cooling effect of green roofs will reduce energy use during the lifetime of the building."[1]

§7 Costs and financial considerations (pp. 29–30): "The build costs when specifying a green roof are undoubtedly greater than a conventional asphalt finish or inverted roof finish. Indicative costs per m^2 of green roof are detailed in Table 10 for various roof types. It is important to recognise that these are guideline prices only and £/m^2 costs will be higher on smaller roofs due to the extra detailing that is required and absence of economies of scale. The table does not include installation costs and these will have to be

[1] **p. 19**.

ascertained to understand the total cost of the roof design as specified. For example, a sedum mat has to be pre-grown and therefore incurs greater production costs than other types of extensive green roof but its installation costs are generally less. Installation costs vary widely and are higher in the UK than in Germany where the green roof market is very competitive and there is a high level of experience among the German construction industry in installing green roofs. Installation cost will fall in the UK as more contractors become experienced in installing green roofs. [...]"

Type of green roof finish	Cost per m^2	Installation issues	Maintenance costs
75 mm substrate sedum plug roof	£25	Slower installation as plants installed separately	Quite high over first 2 years while systems bed in
75 mm substrate sedum hydro seeded	£ 20 −25	Fast installation; can take 2 years for plants to mature	Quite high over first 2 years while systems bed in
75 mm natural/ biodiversity	£ 20 −25	Fast installation; can take many years for plants to establish	Low
20−30 mm sedum mat/blanket	£40 −50	Fast installation; immediate green roof effect	Low−medium

Table 10 (p. 29) Average costs for various types of green roofs.

"It is possible to offset some of the costs of a green roof at the design and specification stage. For example, an inverted roof system requires ballast, either shingle or paving slabs, which can be replaced by a 75 mm substrate based green roof. Ballast paving costs in the region of £10−17 per m^2. If a substrate based biodiverse roof at a cost of £20−25 per m^2 replaces paving at £17 per m^2, then the additional cost of a green roof is reduced to £3−8 per m^2. As green roofs are considered as a source control component of SuDS, savings can be made further down the SuDS management train when a green roof is specified. These savings can be significant if applied to many roofs across large developments, especially as most downstream SuDS solutions such as swales, detention basins and balancing ponds require land grab, whereas green roofs do not.

It is well established that operational costs of a building with a green roof will be less in terms of the amount of cooling required. As the UK climate heats up and the cost of electricity rises, these savings will increase. In Germany, it is estimated that the cost to install a green roof and maintain it over 40 years is about 43 (£29) per m^2, compared to a possible saving of 70 (£48) per m^2 from the reduced maintenance, energy saving, city water fee (saving from storm-water runoff) and increased life."

§6 **Retrofitting (pp. 26–28)** addresses some answers about refurbishing with green roofs, although the information is just a general overview.

Guidelines (3): CIBSE-Chartered Institution of Building Services Engineers, 2013, Guidelines for the Design and Application of Green Roof System

The aim of these guidelines is to highlight effective methods for the planning, design, and implementation of sustainable green roof systems, to increase the awareness of green roof systems and encourage their use on new developments and existing buildings.

Applicability

- Considerations for the planning, design, installation, and maintenance of green roof systems, to provide education and ensure effective application of green roof schemes;
- Certain technical areas of green roof technology (i.e., drainage flow rates, growing medium performance criteria, or waterproofing) are not in the scope of this document and other sources are referenced in the appendixes for addressing these topics.

Structure

The guidelines are divided into eight chapters and eleven appendixes:

1 Introduction: the purpose of the guideline and the benefits of green roofs in urban management and in the public and private sphere

2 Definitions of green roof types (i.e., intensive and extensive), basic construction methods (i.e., built-in and modular green roofs), and considerations for green roofs on new and existing buildings

3 Explanation of the structure of a green roof and references to the German and Japanese standards/guidelines (FLL and Urban Green Tech, respectively)

4 Planning requirements, such as structural loading, accessibility, site conditions, waterproofing, and green building credits/incentives (i.e., BREEAM, LEED, BEAM plus)

5–8 Considerations for the design (i.e., irrigation, drainage, plant selection, sustainable technologies, food production, etc.), construction (i.e., installation, fire and safety precautions, inspections, testing, etc.), maintenance (i.e., procedures, warranties, liability, irrigation, fertilization, etc.), and management (i.e., contractual matters, cost of green roofs, policies, etc.) of green roofs

9 References

Appendix:

1 Lists useful websites and resources

2 Descriptions of different types of green roofs

3–4 List of international technical guidelines, codes, and standards for green roofs

5 Checklist for planning the construction engineering and vegetation technology

6 A sun path diagram for Hong Kong and a shading study

7 Design considerations for pitched green roofs

8–9 Checklists developed by Leila Tolderlund (University of Colorado Denver), regarding maintenance requirements and considerations before starting a green roof project

10–11 Case studies

Characteristics

The guidelines are extensive, providing useful information for the design and installation of a green roof scheme.

About

Fire risk/Fire safety

§**6.3 Fire precautions (pp. 46–47)** gives the following tips:

- "There is evidence suggesting that green roofs can help slow the spread of fire to and from the building through the roof, particularly where the growing medium is saturated. However, plants can present a fire hazard if they are dry."
- "Certain plants should be avoided for resafety reasons, such as plants that are highly flammable, develop large root systems and thus high biomass or are excessively thirsty."
- The integration of fire breaks is recommended.
- "Traditional 'biomass roofs' made of reeds [which are not considered green roofs] may be a high fire risk. In general, fire risk of extensive green roofs is not high because the heat load of burning the drought-tolerant plants is low (burn-resistant) and flammable materials are not

used for green roof components. In fact, in some buildings, extensive green roofs can be used to reduce the risk of fire. Intensive green roofs provide preventative fire protection in the case of sparks and radiating heat. [...] Openings within the green roof (e.g., skylights) need to be installed with vegetation free zone (approximately 500 mm). On larger roof areas, a vegetation-free zone or fire stops (e.g., gravel strip or concrete slabs) should be installed at least every 40m and around the parapet."

Structural safety/Wind design

Wind design is developed in §**5.3 Wind design (pp. 30–32)** and is summarized in the following table:

Section	Content
§**5.3.1 Wind uplift and ballast**	"Wind loads can damage anything built on top of the roof, either during construction or after work has been completed. The following areas on the roof are most affected by the wind pressure uplift and require appropriate protection: • corners: stress levels are very high; • edges: stress levels are high; • the central area: stress levels are low. Wind pressure can vary across a roof, depending on location. At the centre of the roof, a thin layer of soil may be adequate. At perimeters and corners, high winds may necessitate the use of large stone ballast or multiple rows or precast pavers to prevent uplift. A strip of gravel, stones or pavers around the edge can prevent the wind damage; these also act as vegetation barriers. In general, taller structures have a greater risk of wind uplift. Ballasting requirements vary by building height, parapet height and wind design speed. SPRI (2010a) provides a method of designing wind uplift resistance of vegetative roofing systems. In green roof systems, ballast is the weight provided by growing medium, stones or pavers to provide uplift resistance for roofing systems that are not adhered or

Continued

Section	Content
	mechanically attached to the roof deck. The ballast also provides drainage options when the roofing membrane is mechanically attached or fully adhered. When growing medium is installed and protected by vegetation that nominally covers the visible surface of the growing medium, or the growing medium is protected by a system to prevent wind erosion, the weight of the inorganic portion of the growing medium can be considered ballast weight. When modular trays that are filled with growing medium are covered by vegetation that nominally covers the visible surface of the growing medium, or the growing medium is protected by a system to prevent wind erosion, the weight of the inorganic portion of the growing medium can be considered ballast weight. Ballast can also consist of large stones, paver systems or lightweight interlocking paver systems."
§5.3.2 Calculation of wind uplift design pressures	"Low-rise buildings present fewer challenges in regard to positive or negative wind forces. A lower parapet design in a low-rise building may avoid potential air turbulence and help minimize uplift forces. High-rise buildings are exposed to increased wind pressure, resulting in higher positive and negative uplift forces. Increasing the parapet height can be an effective tool in moderating uplift forces. Some green roof systems have been tested for wind uplift resistance. The wind uplift design pressures and wind rating pressures of the systems are determined. A minimum safety factor (e.g., 1.7) should be applied for wind uplift calculations based on appropriate wind uplift design pressures for the local design wind loads. The calculation is based on a dry condition (no water present in the growing medium, retention

Section	Content
	mat, drainage panel, etc.) and without the presence of any vegetation load. Growing medium may be used as a secondary ballast material; that is, it may be used to ballast the loose-laid roofing components above the waterproofing membrane (i.e., drainage panel, retention mat, root barrier and insulation board) but not the membrane itself. When growing medium is used to ballast the loose-laid above-membrane components, the wind uplift calculations should use wind uplift design pressures, a dry condition (no water present in growing medium, retention mat, drainage panel, etc.), no vegetation and a minimum safety factor of 0.85."

Irrigation

The irrigation topic is explained in general terms in §**7.5 Irrigation (p. 57)**, but this topic is not extensively explained.

Maintenance/Cost estimation

In §**7 Maintenance issues (pp. 52–58)** the essential role of maintenance for a green roof to functioning successfully is explained: "All roofs—green or conventional—require regular maintenance, which can determine the success of green roof systems (Urbis Limited, 2007)."

The contents of this section are synthesized in the table below:

Topic	Content
Factors that determine maintenance requirements	Height, microclimate, soil types, soil depth, irrigation, species, and access. In Hong Kong access influence maintenance costs, and green roof design should take this into consideration.
Three stages of maintenance	**(1)** Installation maintenance: several aspects of maintenance and service are needed throughout the first year. It is important to provide a sufficient water supply during the

Continued

Topic	Content
	dry season. Replanting is necessary if there are dying or missing plants after installation and weeds and other unwanted plants need to be removed. **(2)** Development maintenance: it is essential to support the vegetation until total coverage of the roof is achieved. The development maintenance is a lower-intensity equivalent of the installation maintenance. **(3)** Upkeep or ongoing maintenance: after the plants on the green roof have developed, it is crucial to maintain and inspect the roof once or twice a year. Weeds and other unwanted plants on the entire roof, at the perimeters and at the upstands need to be removed. The irrigation system and inspection chambers need to be monitored to ensure that water outlets are free from blockages. For grass and herb vegetation, the organic material has to be removed once a year and grass needs to be pruned regularly. Intensive green roofs require higher maintenance and service throughout the year.
Maintenance procedures	Irrigation, weeding, fertilizing, pruning, and replanting (Some procedures should be scheduled after events, such as floods and storms, while others can be scheduled according to seasonal events)
Waterproofing membrane maintenance	Maintenance and visual inspections of the waterproofing membrane can be complicated by the fact that green roof system completely covers the membrane. [...] Regular maintenance inspections should be schedule as for a standard roof installation, especially just before the warranty period ends. Eventually, after 30–50 years, the membrane will have to be replaced.
Maintenance consideration during the design phase	**(1)** It is important that any budgetary constraints are accounted for. **(2)** The cost of the roof maintenance should form part of the lifecycle cost analysis for

Topic	Content
	the building, allowing the most appropriate green roof specification to be realized. **(3)** Access for equipment and inspections following construction. **(4)** Access for water is very important in most green roofs. Irrigation system, growing medium and plant selection are critical factors determining long-term maintenance requirements and survival of the green roof vegetation under hot, dry conditions; otherwise, vegetation may have to be repeatedly replanted and/or the irrigation system replaced. Access to multi-functional irrigation timers is recommended. **(5)** If an underdrain system is used, provide a clean-out for both inspection and maintenance. There is potential for it to become clogged with growing medium and organic matter that migrates down beneath the plant root zone. The ability to access the underdrain system, flashings, drains, etc., for clean-out is essential. **(6)** All drains must remain free of vegetation and foreign object and permanently accessible.
Fertilization	Reducing or eliminating fertilization of green roofs is recommended, as it can cause pollution in local water courses.

Table 7.1: Maintenance procedures for green roofs.

Maintenance procedures	Intensive: Podium or sky garden	Extensive: Extensive green roofs on existing roof	Notes
Waterproofing inspection	Once a year	Once a year	Inspect for water penetration through the concrete.
Drainage inspection	Once a month	Every 2 months	Inspect drainage outlets.
Litter	Once a week	Once/as necessary	Includes removing litter and emptying bins, depends largely on the

Continued

Maintenance procedures	Intensive: Podium or sky garden	Extensive: Extensive green roofs on existing roof	Notes
			type of green roof, visitor numbers and occasional litter that may fall on it.
Plant health inspections	Six times a year	Twice a year	Check for insect and fungus infestations (particularly relevant during establishment period).
Replacement planting	As necessary	As necessary	
Irrigation	Irrigation 780 L/ m^2 per year	Low to none	Based on 15 L per week (reduced from 25 L due to Hong Kong's high rainfall during summer) amenity planting.
Pruning	Twice a year	N/A	
Lawn mowing and rough grass cutting	Nine times a year Up to three times a year	N/A Up to three times a year	Depends largely on design and turf grass species. Fallow grasses require less cutting than turf grass.
Fertilizing	Once or twice a year	Every 4 −5 years	May be increased for horticultural practices that remove biomass from the system, such as lawn mowing.
Disease and pest control	Four times a year	Four times a year	Regular inspections needed.
Weeding	Nine times a year	Up to three times a year	Weeding for extensive green roofs should hardly be needed if installed it is correctly, self-seeded trees may be a problem in urban fringe areas.

Cost estimation is briefly explained in §**8.3 Costs of a green roof system**, but this topic is not extensively developed.

Guidelines (4): Groundwork South Yorkshire, 2011, The GRO Green Roofs Code

The Green Roof Code represents a summary of the best practices, regarding the design, construction, and maintenance of green roofs. It is a preliminary document, which points out relevant instructions, but for extensive details, regarding the design, construction, and maintenance of green roofs systems, it is necessary to refer to FLL green roof guidelines or other standards.

Applicability

- A preliminary document and overview, useful for designers, contractors, and stakeholders, in providing general information on the design, construction, and maintenance of green roofs.

Structure

This guide is divided into seven parts:

1 Introduction to the Code
2 Introduction to Green Roofs (definitions; types; benefits)
3 Design Considerations and Practical Implications: (Configuration of a Green Roof; Structural Design; Waterproofing; Drainage; Fire; Irrigation; Safety and access)
4 Green Roof Installations (Site preparation and planning; Installation of system components; Installation of perimeter and penetration details)
5 Green Roof Maintenance (General Maintenance Actions; Maintenance Actions by Roof Type)
6 Relevant Complementary Documents: list of other important standards and documents, quoted in the code
7 Glossary

Characteristics

This guide is organized in a clear and complete sequence, from the design and installation of the systems to the continuing maintenance requirements. Although this guide is considered a preliminary document, the provided instructions and explanations are "in-depth" and other standards are referenced in a very detailed index for more instructions (in the initial acknowledgments it is written that GRO recognizes FLL as base guideline for green roofs specifications in the United Kingdom).

About

Fire risk/Fire safety

The fire topic, provided in §**3.5 Fire (p. 21)**, points out some important recommendations, as seen in the following table.

Section	Contents
§**3.5 Fire (p. 21)**	"However, like any vegetation-covered surface consideration must be given to methods preventing the external spread of flame during prolonged periods of drought include the use of temporary irrigation system to prevent drying out. Depending on the system specified the season maintenance requirements can impact on fire risk. Tall grasses and/or wildflower species that die back seasonally leaving large volumes of organic material on the roof can increase fire risk if the correct maintenance regimes are not followed. All green roof systems should be designed to incorporate gravel fire breaks at perimeters and penetrations, i.e., rooflights, soil pipes, rainwater outlets, etc. These fire breaks should be a minimum of 300mm wide and 50mm deep (see 3.5.1.2)."
§**3.5.1 Extensive roofs (p. 21)**	"Extensive roofs do not tend to be irrigated and the fire risk must therefore be mitigated by the specification of the build-up and the incorporation of fire breaks. Ongoing maintenance should take into account any dry spells that could impact on the volume of dry vegetative material on the roof (especially for wildflower system), and appropriate action taken to ensure no significant volume of dry material is left on the roof. This is particularly important on roofs overlooked, where an ignition source such as a burning cigarette could be discarded onto the roof causing a fire."
§**3.5.1.1 Green roof built-up (p. 21)**	"Substrate should have • A depth in excess of 80 mm; • a minimum depth of 80mm; • A maximum of 20% organic matter; • Plants such as succulents retain water within their structure, reducing the risk of drying out."

Section	Contents
§**3.5.1.2 Fire breaks (p. 21)**	• "Non-vegetated strips comprising 20–40mm rounded pebbles or concrete paving stones that should be kept clear of encroaching vegetation during maintenance visits. The width of these strips should be a minimum of 300mm but should be increased to a minimum of 500mm where the strip is providing preventative fire protection such as opening rooflights and vertical walls with windows or opening doors; • On large roofs a 1 meter wide pebble strip should be installed at 40 meters intervals across the roof. Installed in 1 m strips (or 300 mm heights) at 40 m intervals across the roof; • To help maintain the vegetation barrier, a retention angle should be included between the growing medium and the pebble margin/ paving. This area should also be subjected to routine maintenance to remove any invasive plant species."
§**3.5.2 Intensive roofs (p. 21)**	"DIN Standards (the German equivalent to British Standards) have designated 'intensive greening which is irrigated, regularly maintained and has a thick substrate layer' as a 'hard roof'. This implies that it has no greater fire risk than a conventional roof finish. On this basis, GRO believes that intensive roofs should be deemed to meet BS 476-3: 1958 designation EXT. FAA. For extensive and biodiverse green roofs, the manufacturer/supplier should be consulted, as should Agreement or other relevant certification."

At the beginning of this section, it is specified that: "Many green roof systems will provide a positive contribution to the resistance of penetration by fire due to the noncombustible nature of mineral/brick-based growing media. Green roof systems can also provide a positive contribution to resistance of spread of flame provided they are correctly designed, installed, and maintained."

Moreover, intensive green roofs are considered "hard roofs" if they are irrigated, regularly maintained, and have a thick substrate layer (§**3.5.2 Intensive roofs**).

Structural safety/Wind design

§**3.2.1 Wind (p. 18)** specifies that, "Where green roof elements are being used as ballast to provide the roofing system with resistance from wind uplift (i.e., the roofing system is not mechanically fastened or bonded), sufficient weight must be incorporated into the green roof build-up. In these instances, the dry weights of the green roof components must be used to calculate the weight of the green roof system. Subject to wind suction loads, erosion control measures, such as netting, may be required while the planting establishes. With highest wind load areas occurring at roof perimeters, and corners in particular, heavier materials, such as larger ballasting aggregates or paving slabs, should be used to prevent wind scour of the green roof substrate."

§**3.2 Structural Design (pp. 18–19)** lists the standards to refer for structural design:

"A green roof design must comply with all relevant structural design criteria, as per BS EN 1990:2002 'Eurocode—Basis of Structural Design'.

As such, designs must be in accordance with all appropriate Eurocodes, with a notable emphasis on EN 1991—Eurocode 1: Actions on structures.

On projects covered by NHBC warranty, the structural design shall be undertaken in accordance with recognized standards.

Items to be taken into account include

(a) Dead and imposed loads—these should be calculated in accordance with BS EN 1991-1-1, BS EN 1991-1-3, and BS EN 1991-1-4. Where a flat roof is to act as a roof terrace, roof garden, or will have vehicle access, appropriate provision should be made for additional loading conditions. Intensive green roofs should only be used in conjunction with concrete decks.

General Actions:

- Densities, self-weight, imposed loads for buildings (BS EN 1991-1-1:2002);
- Snow loads (BS EN 1991-1-3:2003); and
- Wind actions (BS EN 1991-1-4:2005 +A1:2010 Eurocode 1. Actions on structures. General actions. Wind actions).

The building structure and any loaded roof components (e.g., insulation, waterproofing) must be designed accordingly."

Irrigation

The irrigation topic is briefly addressed in the §**3.6 Irrigation (p. 22)**:

"Irrigation is typically required for the initial establishment of the green roof for a period of 6–8 weeks until established depending on natural rainfall during this time. However, once plant cover is achieved, irrigation can be reduced (for intensive and semi-intensive roofs) or avoided (for most extensive roofs, subject to plant selection).

The more intensive the roof, the more likely it will be that an artificial irrigation system is required.

The requirement for irrigation depends on many factors, particularly

- The planting layer's water demands;
- Water storage capacity of the green roof configuration (e.g., growing medium, drainage layer);
- Local precipitation patterns.

In all instances advice should be sort from the green roof system supplier.

Once established rainfall is the typical source of water for green roof system; however, complementary irrigation options include hoses, sprinklers, overhead irrigation, and automated systems that pump from a storage reservoir.

The establishment of a need for an irrigation system, and the design of an irrigation scheme, should be in accordance with the principles of

- BS 7562-3:1995 Planning, design, and installation of irrigation schemes—Part 3: Guide to irrigation water requirements.

Standards relevant to the design of specific irrigation systems include

- BS EN 15099-1:2007—Irrigation techniques. Remote monitoring & control system;
- BS EN 15097:2006—Irrigation techniques. Localized irrigation hydraulic evaluation;
- BS EN 13742-1:2004—Irrigation techniques. Solid set sprinkler system—selection, design, planning & installation."

Maintenance/Cost estimation

The maintenance topic is addressed in §**5 Green Roof Maintenance (pp. 27–28)**.

§**5.1 General Maintenance Actions** and §**5.2 Maintenance Actions by Roof Types (pp. 18–19)**.

In §**5.1 General Maintenance Actions**, two standards are quoted as guidelines for maintenance actions:

- BS 4428:1989—Code of practice for general landscape operations (excluding hard surfaces);
- BS7370-4:1993—Grounds maintenance—Part 4: Recommendations for maintenance of soft landscape (other than amenity turf).

There are four listed general maintenance actions, summarized in the following table:

Section	Contents
§5.1.1 Irrigation (p. 27)	"There is a reference to the dedicated section (Section 3.6 for further details)."
§5.1.2 Fertilization (p. 27)	"Fertilization is the process by which additional nutrients can be supplied to the plants, enhancing germination, flowering, and resistance to weather extremes. The regularity and type of fertilization requirement will therefore depend on the type of green roof and its plant specification. If using non—slow-release fertilizer, the potential impact on the sewage system must be considered, and the rainwater system isolated from the mains for the first flush if necessary (see Section 5.2 for further details). Intensive and simple intensive roofs are based on a more fertile growing medium and the planting installed will require regular fertilization."
§5.1.3 Plant management (p. 27)	"Undesirable plant species are best avoided by establishing a complete coverage of the desired plant species. Any wind-blown seeds or cuttings should be removed before they have the opportunity to take root (see Section 5.2.2 for exemptions)."
§5.1.4 General Clearance/ removal (p. 27)	"Generally the removal of dead material is desirable as it allows plants the space to develop a greater coverage, improving the finished appearance of the roof, whilst also reducing the risk of fungal disease forming and spreading. However, in some biodiverse applications, removing plant debris could be counterproductive in creating habitat."

There are no references to fire prevention or the prevention of dead of fallen plant, but as result of regular and complete maintenance activities (always recommended for green roofs), these two risks are avoided.

§5.2 Maintenance Actions by Roof Types (p. 28) provides a general overview of the required maintenance actions that need to be performed for different types of green roofs (i.e., extensive, biodiverse, semi-intensive, and intensive systems).

The following table reports the contents of the Section:

Maintenance action	Green roof type			
Classification	§**5.2.1 Extensive roof maintenance** <100 mm low nutrition substrate	§**5.2.2 Biodiverse** very low to low nutrition substrate	§**5.2.3 Semi-intensive** 100 mm–200 mm low to medium nutrition substrate	§**5.2.4 Intensive** 200 mm + medium nutrition substrate and top soil
Irrigation	"Post-establishment, irrigation should not be required for most extensive green roofs, although the water storage capacity of the system and the plants' water demands should be appropriately assessed."	"Typically not required."	"Periodic irrigation is expected, depending upon the plant specification and the climatic and microclimatic conditions prevailing at roof level."	"Regular irrigation is often required, subject to the plant specification and the climatic and microclimatic conditions prevailing at roof level."
Fertilization	"Extensive green roofs typically have low nutrient requirements and are therefore often fertilized on an annual basis, each spring, using a slow-release fertilizer."	"Generally, not required, particularly where indigenous species are being encouraged to replicate native habitats. While a low vegetative density is common, zero vegetation is generally undesirable."	"With a wider range of planting, using a more fertile growing medium, more regular fertilization is required."	"With a wider range of planting, using a more fertile growing medium, more regular fertilization is required."

Continued

Maintenance action	Green roof type			
Plant management	"Removal of undesirable plant species and fallen leaves should take place twice each year."	"A maintenance programme should be drawn up to follow the biodiversity hypothesis, ensuring that no materials are removed from the roof that may adversely affect the biodiversity potential of the roof."	"Removal of undesirable vegetation on the greened area twice yearly."	"The intensive maintenance of lawns, hedges, borders, etc., is required on a regular basis, so as to maintain the roof aesthetics. Undesirable vegetation should be removed from the green areas at least twice yearly. Failed plants in excess of 5% of the plants installed should be replaced."
General	"Drainage outlets (including inspection chambers) and shingle/gravel perimeters to be cleared of vegetation, twice yearly."	"Drainage outlets (with inspection chambers) and gravel/shingle perimeters should be inspected twice yearly and cleared of any living or dead vegetation."	"Drainage outlets (including inspection chambers) and shingle/gravel perimeters to be cleared of vegetation, twice yearly."	"Drainage outlets (including inspection chambers) and shingle/gravel perimeters to be cleared of vegetation, twice yearly. Where excessive substrate settlement has occurred, this should be replenished."

Guidelines (5): State of Victoria through the Department of Environment and Primary Industry, 2014, Growing Green Guide. A guide to green roofs, walls and facades in Melbourne and Victoria, Australia

This is a technical guide that provides advice on the factors that need to be considered to design, construct, and maintain green roofs, walls, and façades.

The suggested audience of this publication is extremely broad (even professionals with no experience with green systems), as the explanations are comprehensive and include simple information. As this document is not standard, it does not include specific data on layers, materials, tests, etc.

Applicability

- Considerations for the design, construction, and maintenance of green roofs, green walls and green façades;
- Targeted at projects based in Melbourne and Victoria, Australia, but it is largely relevant for any geographical area.

Structure

The guide is divided into four sections and one appendix. At the end, there is a glossary, references, and source material section:

S01—Introduction, definition of what green roofs, walls, and façades are, why they are important, the potential benefits of these systems for projects in Melbourne and Victoria, and some frequently asked questions

S02—Summarizes the benefits of green roofs, walls, and façades through references to research that has already been undertaken

S03—Provides technical advice on the design, construction, and maintenance of green roofs, walls, and façades, including general analysis, design and planning recommendations, considerations to be taken before building and installation (i.e., insurance and system warranties), suggestions for the installation process (including the description of the layering), and maintenance suggestions

S04—Case studies from Victoria, Australia: each case study provides a project description, precise maintenance practices, and associated costs

Appendix:

A Technical and chemical specifications of the growing substrates

Characteristics

It is an extensive and quite complete document, which provides qualified technical information for designers and it includes a technical guide. Another helpful feature is that images, schemes, drawings, and pictures are provided, which provides a clearer understanding for the readers.

About

Fire risk/Fire safety

§**2.6 Nearby vegetation (p. 37)**: "The local vegetation adjoining a site can influence design. If creation of habitat for biodiversity is a desired outcome for your green roof, wall or façade, consideration of the surrounding landscape could be useful. However, nearby vegetation can also be a source of weeds or contribute to a fire risk and should be included in site analysis."

§**3.6 Planning, regulations and local lows (p. 50)** suggests referring to local laws and building standards in case of fire risk.

§**5.6 Growing substrate (p. 73):** "A range of mineral and sheet mulches can be used successfully on green roofs but need to be considered carefully. Organic mulches, particularly fine materials, are generally unsuitable for use on green roofs as they can be easily blown off the roof, degrade rapidly, block drains, or create a fire risk in hot dry conditions."

§**8.2 Maintenance tasks (p. 94):** "Maintenance plans must ensure that the vegetation present on a green roof, wall or façade does not create fire hazards. Dead or dry vegetation must be removed as part of regular maintenance. Fire resistance can be increased by installing breaks in the vegetation or using plants with low biomass (such as native grass tussocks with a low amount of leaf matter). Green façades with deciduous or evergreen climbers are unlikely to create a fire risk as long as dry leaf litter falls away cleanly and is removed during routine maintenance. Vegetation on green walls that receives regular irrigation and maintenance does not pose a fire hazard."

Structural safety/Wind design

§**Definitions: Green façade (p. 7)**: "For multi-level façades, particularly at height, wind can create significant problems for plant attachment. In these settings, twining climbers are preferred over plants that adhere directly to the building façade as the twining stems attach strongly around vertical and horizontal supports. Foliage may still be stripped under extreme wind conditions, so foliage type and size should be matched to the level of exposure and likely wind strengths at the site. In general, the higher the planting on a building, the more extreme the growing conditions are likely to be."

§**2.1 Climatic factors on-site (pp. 32–33)**: "Climatic factors will vary with geographic location as well as with the site aspect and height and even from effects of surrounding buildings. It is important to understand the likely climate on site in order to inform decisions about which plant species are suitable for the site. There are no hard and fast rules about what constitutes too much wind or shade or other factors; rather, these are environmental gradients (for example, low wind to high wind) and often the best approach is to estimate the worst case scenario for plant growth that is likely on-site, and design with that in mind."

"Average wind speeds are greater at height than at ground level. Winds may be strong around the edges of buildings, or from the down draft caused by tall buildings. It is necessary to understand the likely wind load that a green roof, wall or façade will be subjected to, so that it can be built to withstand the forces. Wind at high elevation will also influence temperature, and wind has a direct dehydrating effect on vegetation, therefore influencing species selection and irrigation requirements."

§**2.7 Site assessment summary (p. 38)**: Wind is mentioned as information to collect during the site analysis process, when designing for green roofs, green walls, and green façades.

§**3.8.2, 3.8.4, and 3.8.5 Green roof, wall, and façade plant selection (pp. 53–61)**: Wind is mentioned as a factor to consider when selected vegetation in the design of green roofs, green walls, and green façades.

§**5.11 Wind consideration (p. 78)**: "Even on flat roofs, wind uplift may present serious challenges for retention of substrate and plants. Wind damage can be dangerous to people and property and costly to repair. Retention systems may be required to ensure that materials cannot be blown off the roof.

Wind uplift pressure is lowest on the centre of the roof and highest at the edges, around the perimeter and at corners. The higher the building, the greater is the risk. On pitched roofs, the roof peak is also subject to uplift. Minimizing untethered components greatly reduces the risk of damage to the green roof by uplift.

Where possible, waterproofing should be fully adhered to the roof, or mechanically fixed. For waterproofing layers that are not bonded to the roof, the green roof assembly must provide adequate ballast to prevent uplift. Edge treatments are the most critical: the unvegetated zone around the perimeter of a green roof (Chapter 5.13) may require heavy concrete paving slabs rather than loose gravel ballast. The materials used must conform to the design wind load calculated for the specific green roof location. On some green roofs, perimeter balustrades or parapet walls will provide some protection against wind flow. Other treatments, such as jute erosion control netting (Chapter 5.12) or coated wire retaining systems, may be useful. Plant selection may also be used to mitigate the impact of winds. If the tolerance of different species to wind exposure is understood, plantings can be planned so that the lowest-growing, most robust species are planted in the most exposed areas. Taller, less wind tolerant species are installed behind them. The resulting gentle gradation of vegetation heights interrupts, and slows, wind flow over the planting."

§**5.12 Slopes and wind protection (p. 79)**: "Landscapes on pitched roofs are subject to wind forces and gravity affecting the stability and retention of the growing substrate and plants. In many cases slippage can lead to poor plant performance and ultimately green roof failure. For roof pitches of up to 15 degrees, no additional protection is needed, unless there are strong wind issues on the site. Waterproofing must be root-resistant, and covered with a protection mat. A drainage layer is not always required as the roof can drain effectively through gravity. The stability of steeply pitched green roofs is increased by maintaining vegetation cover: provision of irrigation is essential.

For green roofs constructed on pitches greater than 15 degrees, basic protection can be provided through anti-erosion jute netting installed just below the substrate surface to provide some anchorage to plants. This netting breaks down over time, and is used simply to help keep the substrate in place whilst the plants establish. Greater protection can be provided on steeper sites by using a drainage layer with large cells, or cups. The growing substrate fills the cells of the drainage layer, reducing slippage and providing spaces for plant roots to grow, ensuring further stabilization. A filter sheet must be placed beneath the drainage layer to reduce wash-through of fine particles from the substrate.

For green roofs with slopes between 20 and 45 degrees, "honeycomb" webbing, comprising multiple, enlarged drainage cells, can be installed above the drainage and filter sheet layers (see image). This holds the substrate in place, increases stability and reduces slippage. Other specialized structural elements can be incorporated to transfer shear forces into stable, reinforced abutments that form part of the roof structure, or wind barriers can be installed."

Irrigation

§Frequently asked questions: "Is irrigation necessary for a green roof, wall or façade?" (**p. 11**):

> All green walls and most green façades and roofs, require irrigation. It is possible to install a green roof with no irrigation, but this will limit the range of plants that can be used successfully and the potential benefits of the roof (such as summer cooling or aesthetically pleasing views of leafy plants).
>
> Some green façades grown in garden bed settings may not need irrigation.

§**Frequently asked questions: "How much water do you need?" (p. 11):** "There is not one answer, as calculating the water needed to sustain a green roof, wall or façade depends on climate and environmental influences, the design and type of system used, and on the substrate and vegetation characteristics. Explore alternative sources of water for use in irrigation, such as harvested and recycled water, to minimize reliance on potable water. Many green walls rely heavily on irrigation and it is essential to establish that supply can meet demand."

§**2.1 Climatic factors on-site (p. 32):** "Rainfall in Melbourne is generally not sufficient to support a green roof, wall or façade throughout the year. It is important to establish whether rainwater or another water source can be harvested from other areas on-site, and stored to supply an irrigation system. This will avoid or minimize the need to use potable water for irrigation. It is useful to carry out an irrigation water demand analysis, to estimate water needs."

§**3.2 Planning for drainage and irrigation (p. 45)** suggests the use of the storm water runoff for feeding the irrigation and provides information for dimensioning the irrigation system.

§**3.4 Sourcing skill, expertise and information: Irrigation consultant/Hydraulic engineer (pp. 46–47)**:

- "Advises on a suitable irrigation system during the design process based on the proposed substrate depth/volume and water-holding properties;
- Advises on use of recycled/harvested water, pumps, and storage tank volumes and configurations;
- Advises on approaches to sustainable and efficient water management in the context of the whole site (water sensitive urban design);
- Advises on integration of the irrigation system with the rest of the building's water system."

§**5.10 Irrigation (pp. 77–78)**: "An irrigation system is highly advisable on green roofs: to extend plant species selection options, to improve plant growth rates and increase long-term vegetation success [...] In most cases, irrigation design will be heavily influenced by the nature of the water supply resource (for example, harvested vs. potable water) and a water budget should be used to guide not just irrigation but also plant selection. In larger green roof projects, irrigation is best undertaken by a specialist consultant to guide system design, component selection, installation and maintenance.

Table 11 provides an outline of different options for irrigation of green roofs."

Table 11 (p. 78).

Delivery method	Advantages	Disadvantages
Microspray	Low cost, visible, easy to install, reliable	Uneven distribution (plant interception), high water loss (wind, evaporation), foliage wetting (increased disease potential)
Surface drippers/ perforated pipes	Low cost, visible, even delivery of water	Moderate water loss
Subsurface drippers/ perforated pipes	Low cost, moderate efficiency (water delivery to root zone)	Nonvisible (difficult for maintenance), higher potential for damage by people digging
Subsurface capillary	High efficiency	High cost, maintenance and repair is difficult because not visible, "capillary rise" of substrate needed or water will not reach plants
Wicking associated with irrigation in drainage layer	High efficiency, ease of installation	Linked to proprietary systems, "capillary rise" of substrate needed, this may be unsuitable for plant establishment if the water is applied too deep for the plant roots to reach
Hose	Good for domestic application for easily accessed areas, not so good for other areas. Allows monitoring to occur at the same time	Cost (requires someone to be present on site), low water efficiency, foliage wetting, uneven distribution

Maintenance/Cost estimation

§**Definitions (p. 4)**: "Extensive green roofs are lightweight with a shallow layer of growing substrate of less than 200 mm deep, requiring minimal maintenance. Intensive green roofs need more irrigation and maintenance than extensive roofs."

§**3.3 Designing for maintenance (p. 46)**: "Design and planning of a green roof, wall or façade must incorporate an understanding of how the system will be maintained. Buildings owners and property managers need to understand what is involved in maintaining the roof, wall or façade and must be committed to managing it; otherwise the benefits outlined earlier in this document may not be achieved. The systems, or assets, created must not exceed the skills, technologies and resources of those who will be given the responsibility for their maintenance. There are examples of green roofs in Melbourne that have been designed which require complex horticultural management, but the management has fallen to contract staff with no specialist expertise in green roofs, and the landscapes begin to deteriorate. This can result in user complaints and/or the need to replace an asset prematurely, which is inefficient and unsustainable both economically and environmentally. It may be that the green roof, wall or façade provider is best placed to undertake on-going maintenance on a contract basis. The person or team with ultimate responsibility for management of the project/property must be clear about maintenance objectives and their capacity to undertake them with available resources. All design options proposed must be fully evaluated in terms of the maintenance they will entail. Including ongoing maintenance costs is an important part of the design considerations, especially from the point of view of the client, or asset owner. To determine the ongoing maintenance requirements of a green roof, wall or façade consider engaging a consultant or contractor with relevant experience. Advice can be provided on the resources needed to maintain different design options and the likely expenditure needed to maintain the materials used. For large commercial projects, a maintenance impact statement can be provided by the landscape designer. Alternatively, specify maintenance objectives and standards early, so that the designer has these in mind to start with. Chapter 8 provides information on common maintenance tasks and the development of a maintenance plan. Consideration may be needed for renewal or removal of a green roof, wall or façade, especially where a temporary or short-term installation is proposed. The Melbourne Central shopping centre green wall was designed as a temporary installation, however with various interventions its life was extended for some years.

Ultimately it was dismantled due to high maintenance costs for a wall in a position with very limited light. Although roofs, walls and façades can, of course, be designed from the outset to last for decades, in some cases they will be designed for a limited lifespan and therefore the opportunities for renewal or process for removal should be considered in the design stages."

§**3.5 Cost consideration (p. 48)**: "Typical maintenance costs include irrigation water, fertiliser, replacement plants, weeding and pest and disease management. Periodic inspection and maintenance of the site, from the irrigation system to clearing drains to re-tensioning of cables or repair of loose wall fixings, will be needed. Maintenance costs may include access costs if, for instance, an elevated work platform is required."

§**3.6 Planning, regulation and local laws (p. 50)**: "Building and maintenance of green roofs, walls, or façades. Laws will often address issues related to:

- Management and disposal of waste products, such as pruning material generated by maintenance contractors;
- Management of vegetation considered to be a weed or fire risk;
- Management of overhanging vegetation on public land or that which has a negative impact on lighting or traffic signs;
- Drainage over public land;
- Use of elevated working platforms or cranes on public land during construction or maintenance;
- Access at street level for people with a disability—consider encroachment of plants or structures into footpath space;
- Access by emergency services—do not impede the safe egress of occupants in an emergency."

§**5.13 Hard landscape elements (pp. 80–82)**: "Drains and gutters. All drains must be accessible for maintenance, protected from blockage by leaf litter and substrate wash, and housed with inspection chambers, drain covers, filters or strainers. Inspections after construction, following storms and every 3 months are recommended."

§**8 Maintenance (pp. 92–96)** provides information to help with the development of maintenance plan.

A detailed maintenance plan will outline performance standards, tasks to be undertaken, and the resources required to achieve them. Once created, maintenance plans should be reviewed annually to ensure all maintenance needs are being met.

§**8.1 Maintenance planning (p. 93)**: "A maintenance plan should include a clear description of:

- Maintenance objectives—created based on the design intent, or the landscaping or environmental objectives that were the basis for the roof, wall or façade development;
- Performance targets, such as the time frame for complete coverage of an area by plants and foliage;
- Responsibilities of various personnel involved in operating the building, outlining the type, scope, duration of task and occurrence;
- Training requirements (such as working at heights certification) and safety equipment;
- Resources available.

Maintenance planning should also incorporate risk management, with the aim of reducing or eliminating the likelihood of failure that could result in property damage or personal injury.

For large projects, maintenance planning is often based on 'asset management planning' where the whole life of the asset is considered, including design, construction, establishment, operation, maintenance, renewal and demolition/replacement.

For some green roofs, walls or façades, particularly those located on commercial premises, maintenance will be undertaken by someone other than the building owner. A maintenance agreement with the installation company or with a recommended third party may be the most economical way to ensure the best long-term performance of a green roof, wall or façade. If a maintenance contract is used, it is important to be clear about the duration of the maintenance agreement, the scope of maintenance responsibility, and the need for handover at changeover to either new contractors or back to the building owner. A supervisor may be designated to oversee the overall and ongoing management of maintenance activities, and can provide direction to maintenance staff and assess that work has been carried out satisfactorily. The supervisor's role will involve:

- Scheduling of maintenance: flexibility may be needed, as, for instance, additional visits may be necessary to repair damage following extreme weather events;
- Signing maintenance contractors in and out at the start and end of a visit, offering a toolbox talk at the commencement of a visit, and providing keys and providing any specialist equipment that is required;
- Occupational health and safety: ensuring contractors have appropriate certification; scheduling mandatory annual checks (and repair or replacement if necessary of safety anchor points); ensuring conditions are safe for maintenance staff to perform their work; providing safety

signage or other barriers to restrict access during maintenance; and ensuring that staff understand the process for reporting actual or potential hazards;

- Informing other contractors who work on the building about the roof, wall or façade, so that they do not inadvertently damage the asset (for example, water service contractors turning off water for a prolonged period)."

§**8.2 Maintenance tasks (p. 94)**: "Maintenance falls into a number of categories:

(1) Establishment maintenance occurs during the first one to 2 years after installation and is undertaken to fully realize the design intent and outcomes. For vegetation, this includes tasks such as pruning, weed control, and irrigation to ensure healthy and vigorous plant growth;

(2) Routine or recurrent maintenance includes regular works that are undertaken to ensure the roof, wall or façade is maintained to a minimum or required standard of appearance, functionality and safety. For vegetation, this can include tasks such as weeding, pruning, removal of leaf litter and, in some cases, mowing;

(3) Cyclic maintenance is scheduled interventions at less frequent intervals that maintain infrastructure. It includes maintenance of the underlying building structure and of specific components of the green roof, wall or façade system. This may include infrequent pruning or other formative management of woody vegetation, such as coppicing, or annual treatment of decking or other hard landscape elements to maintain their safety and functionality;

(4) Reactive and preventative maintenance is undertaken when some component of the system fails suddenly, or shows signs of imminent failure. Failure may be due to a long-term problem that has gone undetected (such as blocked drains by tree roots), or sudden damage resulting from an extreme weather event (such as stormwater incursion);

(5) Renovation maintenance includes works that change the design intent. This may arise after a change of ownership of a building which instigates a desire for change, through to remediation of a design failure (see The Venny case study). Some typical maintenance activities for green roofs, walls and façades are outlined in the table below. These are intended as a general guide only: each site will have its own specific requirements and some listed will be more relevant and/or specific to a wall, roof or a façade.

Access for maintenance: Roofs that are not generally accessible to the building owner or tenants may need to be accessed by ladder, using ropes and harnesses or other 'fall arrest' systems to work safely. For small wall and façade installations, within a few meters of the ground, a mobile work platform at ground level is probably the most efficient solution for access. However, ropes and roof-mounted attachments may be required for larger areas or higher buildings. Green walls and façades extending more than 10 m above ground level will usually be maintained from a roof-mounted work platform. A hybrid living wall system developed in Adelaide addressed maintenance access with an inbuilt platform at each floor level that allowed safe access for maintenance of the vegetation and for other services such as window cleaning."

CHAPTER 7

Manuals

Manual (1): GRHC—Green Roofs for Healthy Cities, 2014, Green Walls 101: Systems Overview and Design

Green Walls 101: Systems Overview and Design is a condensed introductory green wall manual developed by Green Roofs for Healthy Cities (GRHC), a nonprofit industry association dedicated to the development of living architecture, primarily through the green roof and wall industry.

Applicability

- General manual for green façades, walls, and living walls.

Structure

The manual is divided into nine chapters and three appendixes:

1. An introduction, describing the mission, activities, and courses organized by GRHC;
2. Overview of the history of green walls and the definition of green façades, living walls, and living retaining walls;
3. Benefits of green walls in public and private settings (i.e., esthetic improvement, urban heat island effect, exterior air quality improvement, local job creation, energy efficiency improvement, building structure protection, interior air quality improvement, noise reduction, marketability, increasing biodiversity, health and well-being, urban agriculture improvement, and on-site waste water treatment);
4. Reference to policies (i.e., Germany, Tokyo, and Seattle), standards (i.e., LEED and Living Building Challenge), and research (i.e., Green façade research in Berlin, Building of Physics Berlin-Adlershof; Green wall Field Test Research, Southern Illinois University Edwardsville; Rainwater Interception Capacity of green façades, Vancouver, BC);

Green Roofs, Facades, and Vegetative Systems
ISBN 978-0-12-817694-8
https://doi.org/10.1016/B978-0-12-817694-8.00007-8

5 Considerations for the predesign phase: objectives (reduce heating/cooling systems, improve air quality, produce food, reduce noise, aesthetics), site (building, surrounding environment, climate), and budget (project size, design team cost, system type, building location, cost of labor, availability of materials, timeline, level of maintenance);
6 Description of green façade and living wall's design, construction system (i.e., modular trellis panel, wire rope and cable net, and secondary supporting structures), irrigation system, growing medium, drainage, plant selection, installation, and maintenance;
7 Case studies of green façade and green wall examples;
8 The final exercise for four scenarios;
9 References.
Appendix
I. Green walls and LEED Certifications
II. Sample Green Wall Implementation Timeline
III. Green Walls Resources

Characteristics

The manual classifies green façades and provides descriptions of the systems, installation considerations, maintenance considerations, some technical details, and some useful information for designers for each identified class. Also, in §**7 Case studies (pp. 61–85)**, heterogeneous system types are introduced with some specific information and "lessons learned" for each case study.

Two of the topics investigated (Fire risk/fire safety and Structural safety and wind design) are beyond the scope of GRHC Green Walls 101 manual.

About

Fire risk/Fire safety

N/A.

Structural safety/Wind design

§**5.1.1 Site analysis (p. 37)** takes wind as environmental factor into consideration, but no design rules are given.

Irrigation

For each class of green façade, there is a paragraph "Water Supply and Drip Irrigation":

Green Façades		
Modular trellis panel	Wire rope and cable net	Living walls
§**6.1.3 (p. 47)**: "[...] Nearby structures, building drainage or underground water removal systems may limit a plant's ability to depend on moisture and temperature related stress on the south and west exposures of structures since they are exposed to a greater degree of sun than other building exposures. All of these factors contribute to the need for an irrigation system in order to ensure plant survival. Automatic irrigation systems are the most commonly used system type for climbing plats; in temperate climates alternative irrigation system may be used, such as low flow bubblers and low angle strip spray system. Factors that affect irrigation system design are: plant selection, local water availability and water use, restrictions, onsite water source, building and sun exposure, LEED certification and ecological design goal."		§**6.2.2 (p. 56)**: "[...] In colder climates, outdoor living walls require daily irrigation during the growing season and supplemental water during the winter to avoid desiccation. Indoor living walls will require consistent irrigation year round and may create a humid micro-climate, therefore the HVAC system should be design accordingly. [...] There should be measures in place such as a backflow preventer and water quality monitoring to avoid contaminations by pathogens, especially in indoor setting. Irrigation system can take several forms including: drip irrigation with emitters located above each panel, a channel system carrying water along each row of plants, continuous synthetic moisture fabric. [...] The greatest risk to a Living walls is interruption of water supply. A maintenance contract and remote monitoring of large or highly visible walls is recommended."

Maintenance/Cost estimation

§**5.1 Scope (pp. 35–36)** explains: "In instances where a green wall is successfully installed, the growth vigorous and the maintenance contract in keeping with the design intent, it is in the owner's best interest to continue allocating annual funds for its upkeep. An annual maintenance budget is much less costly in the long run than reparation or replacement of dead plant materials."

<table>
<tr><th colspan="2">Green Façades</th><th></th></tr>
<tr><th>Modular trellis panel</th><th>Wire rope and cable net</th><th>Living walls</th></tr>
<tr><td colspan="2">§6.1.7 Maintenance Requirements (p. 53): "Maintenance is crucial to the success of a project and should be considered in the initial stage of design.
[…] A maintenance plan should address the following:
• Irrigation requirements
• Soil amendments
• Access to plants on the wall
• Plant inspection for disease and pests
• Rates and methods for pruning and training leaders
• Adjustment to underlying system to accommodate growing weight
• Method of accessing the underlying wall structure for repairs (this should include a detailed guide on plant storage and maintenance when removed from wall).
It is important that the owner be aware of the direct relationship between maintenance and a green wall's ability to perform its function."</td><td>§6.2.7 Maintenance Consideration (p. 60): "The success of a living wall is highly dependent on maintenance, more so than green façades.
[…] A maintenance plan should address the following:
• Monitoring schedule
• Method of accessing plants on the wall
• Most local building codes require that all workers have fall safety training before they work heights
• Irrigation requirements
• Plant nutrient requirements
• Soil quality maintenance
• Pruning requirements
• Disease and pest control
• Methods of plant or module replacement
• Adjustments to underlying system to accommodate growing weight
• Methods of accessing to underlying system (e.g., Irrigation lines, solenoids and pumps, if present)."</td></tr>
</table>

§**5.3 Budget Development (p. 38)** declares that "No two green walls are the same and each needs to have the associated costs and benefits calculated individually."

Then, it specifies: "There are many variables that influence the capital and maintenance costs of green walls including:

- Project size
- Design team cost

- System type
- Building location
- Local cost of installation labor
- Local availability of materials
- Project timeline
- Degree to which maintenance is required."

§**5.3.1 Cost Estimation (p. 39)** briefly overviews cost considerations: "The allocated capital and maintenance budget will impact the green wall system selected and will determine the outcome of the design process. [...] Maintenance requirements will vary from system to system, whose cost should be included in the initial cost estimation. It should be noted that green wall façades climbers require maintenance during the establishment period (i.e., fist 2–3 years after implementation). Climbing plants used on green façades often require on-going irrigation, fertilization, mulching, weeding, training, pruning. The supporting structure's hardware very rarely requires maintenance. Only in humid regions will it require and occasional rinse with clear water. This should be carries out according to the manufacture's recommendations. It should be noted that living walls require continuous maintenance of the plant life and system infrastructure. Living walls require intensive maintenance including daily or regular watering, a consistent supply of nutrients to fertilize the plants, pruning and plants replacement."

Manual (2): GRHC—Green Roofs for Healthy Cities, 2011, Advanced Green Roof Maintenance

This document is the official "Participant Manual" of GRHC (Green Roofs for Healthy Cities) for the maintenance of green roofs from the inception of the system's design to the postestablishment period.

Applicability

- Extensive overview of maintenance considerations for green roofs, specifically for the North American green roof industry;
- Suggested to be used by green roof professionals, maintainers, facility managers, and roofing contractors, but it is also a useful tool to inform owners and designers (although it is not a design handbook).

Structure

This manual has twelve chapters and four appendixes:

1 Introduction
2 Introduces design, installation, and maintenance
3 Detailed maintenance planning and budgeting
4 Special considerations for working on a roof: safety, access, and egress
5 Maintenance procedure: during installation; during the plant establishment period; and postestablishment
6 Diagnosing problems/diagnostic tools
7 Irrigation systems (i.e., drip systems, overhead spray systems, hand watering)
8 Contractor approaches
9 Contracting maintenance
10 Warranties
11 Case studies
12 References
Appendix

- Glossary
- Sample Maintenance Plan
- Sample Maintenance Report
- Maintenance Requirements for ANSI Fire and Wind Standards

Characteristics

The manual is very detailed, including a chapter dedicated to maintenance planning and budgeting.

The manual does not provide technical sketches.

About

Fire risk/Fire safety

Appendix D (p. 112) mentions the ANSI/SPRI VF-1 External Fire Design Standard for Vegetative Roofs.

Structural safety/Wind design

Appendix D (pp. 112–113) mentions the ANSI/SPRI RP-14 Wind Design Standard for Vegetative Roofing Systems.

Irrigation

The irrigation topic is outlined in Chapter 7: **Irrigation System (p. 69)**. Several technical aspects about irrigation (i.e., functioning, scope, use, types, etc.) are introduced.

Detailed information, but no schemes or sketches, is provided for the following classification of irrigation systems:

(1) Drip irrigation;

(2) Overhead spray systems;

(3) Hand watering.

§**7.4 Standard Seasonal Maintenance (p. 73)** considers the maintenance of irrigation:

When	Suggests
Winterization	[...] In cold season "the water in the system shall be removed so that it cannot freeze and crack the system. This is commonly done by turning off the system, connecting it to an air compressor and blowing air into each within the system to ensure all the parts are dry. Care must be taken so that the air pressure can blow lines apart. Once all the water has blown out of the system, measures should be taken to ensure that water supply is shut off or disconnected so that no additional water will enter the system."
Spring start-up	"In spring when it is time to being using the irrigation system again, all parts should be reconnected to the water supply (if they have been separated) and the water should be run through the system. It is important to run the system with the end of line (flush line open) if it is a drip system in particular, to clean any debris or particles that may have found their way into the lines. Once the lines have been flushed, the system can be tested. Each zone should be run to make sure that there was no damage from winter; repairs should be made at this time if any damage is present."
During the Year	"The system will need to be adjusted depending on the climate and types of plantings. Typically, a system will run less in the spring and fall and more in the summer. Rain sensor should be installed to ensure that the irrigation system is not used when adequate water is supplied from rainfall. Also, especially with drip systems, it is recommended that the filter be cleaned at every maintenance visit (at least once a season) and the end of the line flush lines are opened and flushed out at least once a season. It is also a good idea to test the system (if it is running consistently) every time you do maintenance on a green roof. Testing the system (running each zone for few minutes) will ensure that there is no damage to any of the parts."

Maintenance/Cost estimation

This manual is entirely dedicated to maintenance of green roofs, so a large amount of information is provided. Being a manual, much of the information is provided with detailed explanations.

Chapter 1: **Introduction (pp. 8–13)** provides an overview of the manual (which is a text for training courses), with preliminary information, such as the purpose of maintenance for green roofs, maintenance challenges, consequences of failing to maintain a green roof, consequences of poor designs that does not consider maintenance, and the installation of green roofs.

Chapter 2: **Design, Installation and Maintenance (pp. 14–25)** address several considerations regarding the following:

- The owner's role in the design process, the green roofs types (extensive, semiintensive, and intensive) with their characteristics;
- Green roofs planting methods (well detailed), design objectives with corresponding green roof types, plants maintenance requirements, and irrigation requirements;
- Maintenance design issues, like access to the roof, to water, to electricity sources, presence of walkways, worker safety, etc.

Chapter 3: **Detailed Maintenance Plan and Budgeting (pp. 26–33)** lists the factors to consider when developing a maintenance plan, which are summarized in the following table:

Topic	List of factors to be considered
General Information	• A statement outlining the design objectives of the green roof; • An overall site map with a building plan indicating access routes to the roof, staging areas and parking areas from maintenance vehicles; • An overall roof plan indicating the green roof area, roof access points, water and power access location; • Details, plans, specifications, and shop drawings that describe the components of the green roof assembly; • A detailed planting plan of the roof showing the intended location of different plant species; • A plan of how and when plant assessments should be performed; • A detailed irrigation plan (if applicable) indicating piping layouts, valve and controller locations; • Applicable warranties and contact information for warranty providers;

Topic	List of factors to be considered
	• A contact list for all parties involved in the design, implementation and maintenance of the green roof; • A list of emergency phone numbers of individuals who must be informed in the event of emergency.
Immediate Establishment Care	• A plan for watering immediately following planting; • Anticipated frequency of irrigation during the establishment period; • A schedule for weeding during immediate establishment.
Vegetation Free Zones and Roofing Components	• Frequency of inspection for drains and other roof components, such as flashings, that may fall under a maintenance contractor's scope of work; • How and when to clean entrances, planter edges, paths, etc., and where debris should be disposed of.
Plant Care	• Identification of plant species that may require unique growing media amendments; • Typical lifespan of various species used on the roof; • How to identify common regional weed species and techniques for removing them without promoting further growth; • Best management practices for ensuring weed seeds are not transferred onto the roof during regular maintenance; • When to prepare for the onset of weeds—typically the dormant season; • Anticipated frequency for weeding to be conducted during establishment and postestablishment periods.
Water on the Roof	• Type of irrigation system; • Frequency of irrigation: (1) seasonal variables; (2) moisture sensor inputs; (3) guidelines for adjusting water applications during excessively dry and wet periods; • Built-in irrigation: (1) spring start-up instructions; (2) fall shut-down instructions; (3) Ssystem testing including frequency of testing; (4) backflow prevention maintenance and testing; (5) controller updates and circuit valve testing; (6) piping systems (availability of spare parts for repairs including splice kits, connectors, clamps, primers and cements, spare sprinklers heads or drip line, etc.);

Continued

Topic	List of factors to be considered
	• Manual irrigation: access to water supply: (1) Oon-roof hose bibs; (2) Oon-roof quick coupler valves; (3) backflow prevention devices per local codes; • Water features: specific maintenance issues involving ponds or fountains.
Maintenance of Furniture, Paths, Decking, Planters, and Railings	• Recommended products for upkeep of planter edges, furniture, decking, etc., and products to avoid due to potential chemical effect incompatibility with other components of the roof assembly; • Specific recommendations for which types of tools to use and which to avoid; • On-site storage for maintenance tools, ladder, hoses, security harnesses, and other equipment.
Mechanical Equipment	• Frequency for checking areas surrounding mechanical equipment for vegetation and debris; • Where are visible problems with this equipment to be reported?
Seasonal Recommendations	• Recommendations for or against growing media amendments, such as adding mulching materials; • Frequency of removing biomass if necessary.
Fertilizers	• Fertilization schedule for all areas of the roof; • Type of fertilizer to be used including specific product and brand names; • Particular applications for specific plant needs and seasonal color; • Any local restriction for using fertilizers.
Pruning	• Schedule and instructions for pruning and dead heading specific species; • On-site or off-site composting for clippings and dead leaves.
Guying	• Monitoring plants to correct leaning conditions; • Strategies and requirements for securement including specific materials to be used; • Location of guy wires.
Erosion Control	• Intended depth of the growing medium for the entire roof and different sections; • Instructions for measuring the depth of the growing medium; • Recommended steps to prevent possible erosions;

Topic	List of factors to be considered
	• Maintaining existing erosion control measures on the roof: (1) hydromulches—main function is to help with germination; (2) Eerosion control mats/blankets; (3) tackifiers—keep growing media in place; (4) depends on distance from parapet wall—when erosion is constantly happening—move setbacks? • Growing media depth: (1) intended depth of the growing medium for the entire roof; (2) instructions for measuring the depth of the growing medium; (3) recommended steps to prevent possible erosions; (4) augmenting existing media where it becomes deficient.
Pesticides (if are part of the green roof maintenance strategy)	• Integrated pest management program; • Whether an applicator license is required; • Type of pesticide to be used; • Schedules and sensitivities to building occupants and owners.
Herbicides (if are part of the green roof maintenance strategy)	• Whether an applicator license is required; • Type of herbicide to be used; • Schedules and sensitivities to building occupants and owners.

Chapter 3: **Detailed Maintenance Planning and Budgeting (pp. 26–33)** also lists the main factors that can influence maintenance budgets:

- Type of green roof
- Type of system
- Initial coverage
- Size of green roof
- Design complexity
- Type and design of roof assembly
- Drainage outlets
- Type of plants
- Type and depth of growing medium
- Type of irrigation
- Access to water
- Easy access to roof and inspection points
- Storage maintenance equipment
- Site and climatic conditions

- Expectations for level of upkeep
- Natural events
- Growing media testing
- Cost of labor
- Use of roof and visibility

Chapter 4: **Special Considerations for Working on a Roof (pp. 34–36)** frames the topics of safety and access/egress.

Chapter 5: **Maintenance Procedures (pp. 37–59)** provides specific and operative information on common procedures that a green roof maintenance provider may be expected to perform in a maintenance plan.

The maintenance procedures are classified according to their time of occurrence:

- During installation;
- During the plant establishment period;
- Postestablishment.

Then, in-depth information is provided for several topics maintenance operations: foot traffic, weeding, irrigating, fertilizing, plants and growing medium care, pest management, drain check/cleaning, regular inspections, inspections after extreme rain events, and other practical activities on the green roof.

Chapter 6: **Diagnosing Problems/Diagnostic Tools (pp. 60–68)** explains how to respond in the case of leaks from the roof deck and how to control the quality of the growing media (description of the growing media testing).

Chapter 8: **Contractor Approaches (pp. 75–78)** explains how to choose a contractor from a client's point of view.

Chapter 9: **Contracting Maintenance (pp. 79–81)** outlines the basics of maintenance contracts:

(1) Maintenance program, as explained in Chapter 3 with a schedule of the routine inspections;
(2) A budget;
(3) Performance goals and expectations;
(4) Clear channels of communication between all relevant parties;
(5) Protocols for documenting all maintenance visits, procurers, warranty claims, and records;
(6) Inspection protocol following severe weather events;
(7) Procedure for reporting issues resulting from other trades.

Chapter 10: **Warranties (pp. 82–86)** explains the warranty types for green roofs, the duration, and the items not covered by warranties, underling the importance of record keeping.

Chapter 11: **Case Studies (pp. 87–100)** summarizes five case studies, from Canada or northern United States, and examines maintenance practices for the following topics: watering/irrigation, fertilizer, pruning, guying, erosion control, weed control, pesticides/herbicides, schedule, testing, and access.

CHAPTER 8

Building Codes

Building Code (1): International Code Council, IBC—International Building Code 2018

The International Building Code is an effective building code, which is widely adopted in the United States. It was created with a view to ensure that minimum requirements are met for a number of topics, including "safety, public health, and general welfare through structural strength, means of egress facilities, stability, sanitation, adequate light and ventilation, energy conservation, and safety to life and property from fire and other hazards, and provide a reasonable level of safety to fire fighters and emergency responders during emergency operations."[1]

It is fully compatible with the International Code Council's family of 14 codes, which include the International Existing Building Code, International Green Construction Code, International Fire Code, International Property Maintenance Code, etc.

Applicability

No requirements for green roofs or façades are in the scope of this document, but it is briefly mentioned that the load of green roofs must be considered when conducting calculations for structural requirements.

Characteristics

This is an extensive code with a view to ensure that a certain level of safety and comfort is met in all new building projects in the United States. As this document must focus on all aspects of building projects, the specificity of green system requirements is outside of the scope of the topics discussed.

[1] 2018 IBC, [A] 101.3 Intent.

Green Roofs, Facades, and Vegetative Systems
ISBN 978-0-12-817694-8
https://doi.org/10.1016/B978-0-12-817694-8.00008-X

About

Fire risk/Fire safety

Fire prevention is detailed in the several chapters. Green roofs systems are mentioned in the following sections:

IBC Chapter/Section	Subject
§307 high-hazard group H	Hazardous materials and group H requirements **(vegetation is not mentioned in the list of hazardous materials)**
Chapter 7	Fire-resistant rated construction (Fire and smoke protection features in the International Fire Code)
Chapter 8	Interior finish
Chapter 9	Fire protection and life safety systems
Chapter 10	Means of egress
Chapter 27	Standby and emergency power
§3103	Temporary structures

- In §**1507.16 Vegetative roofs, roof gardens and landscaped roofs** it is declared that: "Vegetative roofs, roof gardens and landscape roofs shall comply with the requirements of this chapter 1607.13.3 and the *International Fire Code*."
- In **[BF 1507.16.1] Structural Fire Resistance** it is written: "The structural frame and roof construction supporting the loads imposed on the roof by vegetative roof, roof garden or landscape roofs shall comply with the requirements of Table 601."
- In **1607.13.3 Occupiable roofs** and **1607.13.3.1 Vegetative and landscape roofs** the calculation of load of green roofs is specified.

As result, it seems that vegetative roofs systems should meet the same fire classification of all other roof assembly materials.

Structural safety/Wind design

N/A.

Irrigation

N/A.

Maintenance/Cost estimation

N/A.

Building Code (2) International Code Council, IGCC—International Green Construction Code 2015

International Green Construction Code is an important building code, which is widely adopted in the United States. It was created with a view to reduce the energy use and carbon footprint of new and existing buildings. The International Green Construction Code (IGCC) provides code regulations with specific requirements to "promote safe and sustainable construction."

It is fully compatible with the International Code Council's family of 14 codes, which include the International Existing Building Code, International Green Construction Code, International Fire Code, International Property Maintenance Code, etc.

Characteristics

General specifications for vegetative green roofs are provided.

Although limited requirements are provided for the four topics of study (i.e., fire, wind, irrigation, and maintenance), there are specifications for classifying an acceptable vegetative roof system.

About

"Vegetative Roof" is defined in §**Definitions** as "an assembly of interacting components designed to waterproof and normally insulate a building's top surface that includes, by design, vegetation and related landscaping elements."

In §**405.1.5 Engineered growing media**, vegetative roofs are briefly mentioned.

§**408.3 Roof surfaces** specifies: "no less than 75 percent of the roof surfaces of buildings and covered parking located in climate zones 1 through 3, as established in the *International Energy Conservation Code*, shall be a roof complying with §**408.3.1** [... or] a vegetative roof complying with §**408.3.2**."

§**408.3.2 Vegetative roofs** provides the following specifications for green roofs:

- All plantings shall be selected based on their hardiness zone classifications in accordance with USDA MP1475 and shall be capable of withstanding the climate conditions of the jurisdiction and the micro climate conditions of the building site including, but not limited to, wind, precipitation and temperature. Planting density shall provide foliage coverage, in the warm months, of not less than 80 percent within 2 years of the date of installation unless a different time period is established in the approved design. Plants shall be distributed to meet the coverage requirements. Invasive plant species shall not be planted.

- The soil medium shall be designed for the physical conditions and local climate to support the plants and shall consist of nonsynthetic materials. The planting design shall include measures to protect the engineered soil medium until the plants are established. Protection measures include, but are not limited to, installation of pregrown vegetated mats or modules, tackifying agents, fiber blankets and reinforcing mesh. The maximum wet weight and water holding capacity of a soil medium shall be determined in accordance with ASTM E2399.
- Where access to the building facades is provided from locations on the perimeter of the roof, nonvegetated buffers adequate to support associated equipment and to protect the roof shall be provided.
- Nonvegetated clearances as required for fire classification of vegetative roof systems shall be provided in accordance with the *International Fire Code*.
- Plantings shall be capable of being managed to maintain the function of the vegetative roof as provided in the documents required by §**903.1.**
- Installation of plantings shall be in accordance with the roof covering manufacturer's installation instructions and shall not diminish the weather protective properties of the roof covering.

Fire risk/Fire safety

§**408.3.2.4 Vegetative roofs** specifies: "Non vegetated clearances as required for fire classification of vegetative roof systems shall be provided in accordance with the *International Fire Code*."

Structural safety/Wind design

N/A.

Irrigation

N/A.

Maintenance/Cost estimation

In Table 902.1 and Table B101.1 vegetative roofs are briefly mentioned:

Table B101.1 Post-occupancy commissioning reporting.

Construction or system requiring verification	Method	Occurence	Section/ reference standard
Chapter 4: Site development and land use			
Stormwater management system operation	Field inspection and report	See maintenance plan	§403.1
Erosion and sediment control	Field inspection and report	See maintenance plan	§405.1.1
Hardscape and shading provided by structures and vegetation	Field inspection and report	See maintenance plan	§408.2
Vegetative roofs	Field inspection and report	See maintenance plan	§408.3.2

Table 902.1 Commissioning and inspection plan requirements.

Construction or system requiring verification	Method	Occurrence	Section/ reference standard
Chapter 4: Site development and land use			
Stormwater management system operation	Field inspection and report	During construction	§403.1
Erosion and sediment control	Field inspection and report	During construction activities	§405.1.1
Hardscape and shading provided by structures and vegetation	Field inspection and report	During construction and installation	§408.2
Vegetative roofs	Field inspection and report	During installation of protective membranes, base materials, soils, and vegetation	§408.3.2

Building Code (3) International Code Council, IEBC—International Existing Building Code, 2012

International Existing Building Code is an effective building code created to ensure that minimum requirements are met when repairs, alterations, additions, or occupancy changes are made to existing buildings. It encourages the use (and reuse) of existing buildings, while guaranteeing that reasonable upgrades/improvements are made.

Applicability

No requirements for green roofs or façades are in the scope of this document.

This code provides requirements to ensure that a certain level of upgrades and improvements are made for existing building projects, but the specificity of green system requirements is outside of the scope of the topics discussed.

About

General roof refurbishment/replacement is discussed in §706 Reroofing, but there is no reference to green roofs/façades.

Fire risk/Fire safety

Fire protection is discussed in several sections (i.e., §405, §804, §904, §1203, §1504, §1509, Resource A: Guidelines on Fire Ratings of Archaic Materials and Assemblies, etc.), but there is no reference to green roofs/façades.

Structural safety/Wind design

Wind is discussed Appendix C: Guidelines for Wind Retrofit of Existing Buildings ("Prescriptive methods for partial structural retrofit of an existing building to increase its resistance to out-of-plane wind loads"), but there is no reference to green roofs/façades.

Irrigation

N/A.

Maintenance/Cost estimation

N/A.

CHAPTER 9

Safety Codes

Safety Code (1): NFPA-National Fire Protection Association 2015, 101: life safety code

This code addresses the construction, safety, and occupancy measures necessary to minimize the risks from the effects of fire, including smoke, heat, and toxic gases created. It establishes the minimum criteria for the design of egress facilities to allow occupants to promptly move out of the building or into safe areas.

Applicability

- Fire safety measures for new and existing buildings and structures, vehicles, and vessels, but green systems are outside of the scope of this document.

Structure

The standard is divided in forty three chapters and three annexes:

1. Scope, purpose, applications, units, and enforcement of this code
2. List of referenced publications
3. General and NFPA official definitions
4. Description of the fire safety goals and objectives, means of egress, authorities, technical assistance, historic buildings, maintenance, inspection, tests, and emergency action plans
5. Description of performance-based options
6. Classification of occupancies and the hazards of the building contents
7. Description of the means of egress (exits, interior finish in exit enclosures, walking surfaces in means of egress, components, measures, characteristics of the doors, characteristic of the stairways, illumination, ventilation, fire compartments, fire barriers, ramps, accessibility, signage)

Green Roofs, Facades, and Vegetative Systems
ISBN 978-0-12-817694-8
https://doi.org/10.1016/B978-0-12-817694-8.00009-1

8. Description of features of fire protection
9. Description of building service and fire protection equipment (fire detection, alarm, communication systems, sprinklers)
10. Description of interior finishes, contents, and furnishings
11–42. Special considerations for different types of buildings (i.e., special structures, high-rise buildings, health care facilities, 1–2 family dwellings, hotels, etc.)
43. Special considerations for building rehabilitation

Annex:

A. Explanatory material
B. Supplemental evacuation equipment
C. Informational references

This code provides information for life safety of different types of occupancies, but the specificity of safety measures for green systems is outside of the scope of the document.

About

Fire risk/Fire safety

The fire topic is addressed in several sections, but the scope of the document involves egresses, stairs, fire escapes, ramps, safe areas, etc., but not green roofs.

Structural safety/Wind design

N/A.

Irrigation

N/A.

Maintenance/Cost estimation

N/A.

Safety Code (2) NFPA-National Fire Protection Association 2015, 5000: Building Construction and Safety Code

This is the first model building code developed using the full open consensus-based procedures accredited by the American National Standards Institute. It provides design regulations to safeguard life, health, property, and public welfare and to minimize injuries by regulating and controlling design, construction, quality of materials, use and occupancy, location, maintenance, and permits for buildings and structures. The intent of this code is to reduce the probability of injury or death from fire, structural failure, or building use by setting a clear set of goals and objectives. In addition to fire, NFPA 5000 provides requirements for other design provisions, including natural hazards, accessibility, and building interior environmental issues.

Applicability

Measures for the construction, alteration, repair, use and occupancy, maintenance, and demolition of all new/existing/moved buildings and structures, but limited reference to green systems.

Structure

The standard is divided in fifty five chapters and eight annexes:

1. The scope, purpose, applications, units, and enforcement of the code
2. List of referenced publications
3. Lists general and NFPA official definitions
4. Describes the fire safety goals and objectives, accessibility, barrier-free use, energy efficiency, egress, roof access, etc.
5. Describes performance-based options
6. Classification of occupancies and the hazards of the building contents

7–55. Describes considerations for all aspects of buildings, taking into account the primary function (high-rise buildings, hotels, residencies, etc.), different building components (i.e., roof, stairways, mechanical system, foundation, etc.), materials (i.e., fire-resistive materials, hazardous contents, steel parts, wood materials), and other special situations (i.e., safe construction techniques, accessibility, buildings being rehabilitated, and designing a structure for dead, live, roof, snow, wind, earthquake, lateral soil, flood, and ice loads)

Annex

A. Explanatory materials
B. Vermin proofing
C. Flood-resistant construction
D. Construction types and enhanced and fire compartment requirements
E. Supplement evacuation equipment
F. In-building radio systems

G. Green construction

H. Informational references

This code provides information to reduce the probability of injury or death from fire, structural failure, or building use, but the specificity of safety measures for green systems are outside of the scope of the document.

About

Fire risk/Fire safety

§**7.2.3.2.12 Exterior Nonbearing Walls**: Nonbearing exterior walls shall be constructed of noncombustible materials, limited combustible materials, or materials specified in §**7.2.3.2.12.1** or §**7.2.3.2.12.2**.

§**7.2.5.6.7 Exterior walls**: Exterior walls having a required fire resistance rating of 2 h or less shall be permitted to be constructed with any of the following materials:

1. Noncombustible material shall be permitted.
2. Limited-combustible material shall be permitted.
3. Fire retardant—treated wood shall be permitted. Approved fire retardant—treated wood framing shall be permitted within the assembly of exterior walls having a horizontal separation of not less than 60 in. (1525 mm), provided that the fire resistance rating is maintained and the exposed outer and inner faces of such walls are constructed of limited-combustible or non-combustible materials.
4. Cross-laminated timber complying with 45.5.7 shall be permitted provided the exterior surface of the cross-laminated timber is protected by one of the following:
   - **a.** Fire retardant—treated wood not less than 15/32 in. (12 mm) thick
   - **b.** Gypsum board not less than 1/2 in. (12.7 mm) thick
   - **c.** Noncombustible material.

§**7.2.5.6.7 Exterior nonbearing walls**: Exterior nonbearing walls shall be tested in accordance with, and meeting the conditions and acceptance of, NFPA 285, *Standard Fire Test Method for Evaluation of Fire Propagation Characteristics of Exterior Non-Load-Bearing Wall Assemblies Containing Combustible Components.*

§Chapter 38**: Roof Assemblies and Roof Structures (p. 382)** provides detailed roof information, but green roofs are not mentioned.

Structural safety/Wind design

N/A.

Irrigation

N/A.

Maintenance/Cost estimation

N/A.

CHAPTER 10

City Codes

City Code (1) Toronto Municipal Code, 2013, Chapter 492: green roofs

This document is the chapter of the Toronto Municipal Code dedicated to green roofs. It includes some technical information about green roofs construction and includes references to several other standards for further information (i.e., Toronto Green Roof Construction Standard, ASTM E2397.05—Standard Practice for the Termination of Dead Loads and Live Loads Associated with Green Roof Systems, CSA—S478—95 Guideline on Durability in Buildings, etc.).

It is important to underline that, in Toronto, all buildings or building additions built after January 30, 2010 require a green roof in some form, unless permission is granted otherwise.

Applicability

- Requirements for green roofs, according to the Toronto Municipal Code.

Structure

The Chapter is divided into nine articles:

1 Definitions of the terms used in the code
2 Requirements for green roofs: dimensions, maintenance, vegetation health, permits, and exemptions
3 Information about applications: approval, fees, and charges
4 Description of the Toronto Green Roof Construction Standards, in relation to the Ontario Building Code: construction features, slopes, parapets, wind uplift, fire safety, waterproofing, drainage, water retention, vegetation performance, plant selection, irrigation, and maintenance plan
5 Information about the approval of applications, exemptions, or variation of coverage requirements for a green roof by the Chief Planner
6 Description of the Green Roof Technical Advisory Group

Green Roofs, Facades, and Vegetative Systems
ISBN 978-0-12-817694-8
https://doi.org/10.1016/B978-0-12-817694-8.00010-8

7 Indicates that changes to the technical standards in future editions could occur
8 Indicates that the Chief Building Official shall periodically develop and publish green roof construction guidelines and best practices to assist designers
9 Offences and penalties.

Characteristics

As part of the Toronto Municipal Code, this document has limited technical or functional content for general green roofs, as the scope is exclusively for green roofs built in Toronto.

About

Fire risk/Fire safety

§492-6 Application for approval (p. 9): "The 'Green Roof Declaration Form' may require information including, but not limited to: structural design; intended use of the roof and whether or not it will be accessible to the public; and fire safety provisions."

§492-9.F Toronto Green Roof Construction Standard: mandatory provisions (p. 11): "Where roof penetrations, intersecting walls, parapets, upturns or mechanical equipment are clad with combustible materials the design shall include a vegetation-free border zone abutting such features and the vegetation-free border shall be equal to the vegetation height at maturity but in no case be less than 0.5 meters."

Structural safety/Wind design

§492-9.E Toronto Green Roof Construction Standard: mandatory provisions (p. 11): "The applicant shall provide a report, stamped by an engineer, providing wind uplift pressures being designed for (including a description of how the pressures were determined), and describing how the design addresses these pressures."

Irrigation

§492-9.M Toronto Green Roof Construction Standard: mandatory provisions (p. 13): "Adequate measures shall be provided to permit irrigation necessary to initiate and sustain the vegetation during the service life of the green roof."

Maintenance/Cost estimation

§492-3 Maintenance of green roofs and health of vegetation (p. 8): "Every green roof required to be constructed pursuant to this chapter shall be maintained in accordance with the maintenance plan required in the Toronto Green Roof Construction Standard."

§492-9.N Toronto Green Roof Construction Standard: mandatory provisions (p. 13): "The applicant shall develop a maintenance plan for the green roof as per CSA-S478-95 'Guideline on Durability in Buildings' which shall define programs of routine maintenance and inspection sufficient to ensure that the green roof components perform their required functions for the duration of their design service lives.

The maintenance plan shall address the requirements of the specified growth media and vegetation for vegetation survival.

The maintenance plan shall address re-planting, in the event that re-planting should become necessary, and assure that complete coverage at canopy level is achieved within three growing seasons and maintained for the service life of the green roof.

The maintenance plan shall be submitted with the application for a permit for a green roof."

City Code (2) Toronto Green Roof Construction Standard-TGRCS, supplementary guidelines

This document is a municipal standard that provides the minimum requirements, recommendations, and considerations for green roofs in Toronto, while also meeting the Ontario Building Code (OBC) requirements.

Applicability

- Requirements for green roofs, specifically in Toronto.

Structure

This document consists of an introduction and six sections:

1. Definitions and components of green roofs
2. Structural considerations
3. Safety considerations (i.e., fire and occupancy)
4. Waterproofing of green roofs
5. Quality assurance

Characteristics

These supplementary guidelines represent a good example of a document that can fill gaps in green roof recommendations of city standards with information from broader standards and building codes. For example, the OBC (Ontario Building Code) does not strictly recognize all green roof components, so this document provides references to other sources (i.e., FM Data Sheet 1-35, Green Roof Systems, 2.2.16, and FLL), which are internationally accepted as "best practices" the topics missing from the OBC. That being said, although the information in this document is useful for all green roofs, it is developed exclusively for green roof systems in Toronto.

About

Fire risk/Fire safety

The fire topic is explained in §**3.1 Fire Safety (pp. 7–8)**:

"OBC Requirements:

OFC 2.4.2.1	'Combustible materials should not be stored on a roof or adjacent to any building so as to create a fire hazard to the building or its occupants.'
Division B 3.1.15.1 and 3.1.15.2	Every roof covering should have a classification determined in accordance with CAN/ULC-S107-M, 'Fire Tests of Roof Coverings.'
Division B 11.3.3.2	Where existing... roof assemblies are substantially removed in an existing building and new interior walls, ceilings or floor assemblies are installed, structural and re resistance elements should be constructed in compliance with the requirements of the other Parts.

Green roof coverings (including plant material) may not comply with the 'Fire Test of Roof Coverings' standard, particularly if vegetation is dormant. At this point in time, there is no widely accepted testing method developed for Green Roofs.

City of Toronto By-law No.583-2009 § 492-9 F.

F. Fire Safety

Where roof penetrations, intersecting walls, parapets, upturns or mechanical equipment are clad with combustible materials, the design shall include a vegetation-free border zone abutting such features and the vegetation-free border shall be equal to the vegetation height at maturity but in no case be less than 0.5 m.

Best Practices

There is no approved fire testing methodology for green roofs; thus it is not possible to classify green roofs in terms of fire resistance. Factors that should be considered include the following:

- Avoiding vegetation that presents excessive fire risk;
- Keeping the vegetation away from walls, parapets, mechanical equipment;
- Providing noncombustible cladding at areas exposed to the green roof;
- Appropriate location of natural gas line supply;
- A consideration of potential impact of burning vegetation on paths of egress; and
- Requirements for firefighting (portable extinguishers and adequate water supply and/or fire hoses).

The "FM Global Property Loss Prevention Data Sheet 1-35: Green Roof Systems" provides specific recommendations to limit risks related to fire. It considers a green roof as noncombustible with regard to interior fire exposure if it is installed over a concrete deck. The FLL guideline accepts the guidelines dictated by the Association of Urban Construction and Housing in Germany (ARGEBAU), which considers the green roof system to have enough fire resistance if the green roof assembly meets the following criteria:

- The soil substrate has a minimum depth of 30 mm
- The vegetation planted constitutes a 'low fire load'
- There is a distance between the green roof area and any penetration in the roof of at least 50 cm.

The FM document, Data sheet 1-35 2.2.14, also establishes a minimum 50 cm nonvegetated border zones as separation from any penetration in the roof.

In general, design should provide solutions which achieve compliance with OBC Division B and its objectives and functional statements. The FM document (Section 2.2.14) recommends avoiding vegetation types with high risk of seasonal dormancy such as grasses and mosses and limiting the vegetation height to a maximum of 0.9 m.

FM also recommends partitioning the roof area into sections no larger than 1450 m^2 separated by a vegetation free border zone of 0.9 m.

The maintenance program provided with the green roof design should include annual removal of dead vegetation that does not promptly compost and may present a fire risk."

Structural safety/Wind design

The wind topic is explained in §**2.4 Wind uplift (pp. 6–7)**:

"OBC Requirements:

Wind loads

Division B 4.1.7.1.(4)	Wind loads are to be calculated using a reference velocity pressure for 'a probability of being exceeded in any one year of 1-in-50'.
Division B Part 5 5.1.4.1	Building materials, components and assemblies that separate dissimilar environments or are exposed to the exterior shall be designed and constructed to provide sufficient capacity and integrity to resist or accommodate wind up-lift imposed...

City of Toronto By-law No.583-2009 §492-9 E.

E. Wind Uplift

The Applicant shall provide a report, stamped by an engineer, providing wind uplift pressures being designed for (including a description of how the pressures were determined), and describing how the design addresses these pressures.

Best Practices

Division B-Part 4 of the OBC requires that the design resist blow-off by the wind. There are practices in the industry, such as loose-laid, protected membrane roofing systems, which are accepted despite limited design information regarding their quantified wind resistance. Green roofs, however, pose a distinct risk of wind uplift.

FM

The FM Global Property Loss Prevention Data Sheet 1-35 Section 2.2.14 recommends a minimum parapet height above the growth media and vegetation free border zones as follows:

Building Height	Parapet height (above growth media)	Vegetation free/border zone
≤46 m	150 mm	0.5 m
>46	750 mm	≥0.9 m

FLL

FLL provides a list of specific recommendations related to securement with the intention of providing a system that can withstand a load of1.5 times the positive or negative pressure, as prescribed by DIN 1055 standard, Part 1. Similarly, a prescriptive securement process is defined in the FM Global Property Loss Prevention Data Sheet, 1-35, Green Roof Systems, 2.2.3.3 Wind Uplift, Roof Ballast and Safety Factor. The following summarizes the FM recommendations:

- Use a mechanically attached or fully adherent roof membrane system based on appropriate wind uplift design pressures.
- Use of growth media as ballast against wind uplift for the roofing membrane and other waterproofing elements only where a uniform depth of 200 mm or more is provided; a minimum safety factor of 1.7 (or as defined by the green roof designer) should be used for wind uplift calculations.
- Use a minimum of 76 mm of stone ballasts in the ballasted areas in the non-vegetated borders, unless greater depths are recommended either by the manufacturer or installer.
- Growth media may be used as a secondary ballast material to ballast the loose-laid roofing components above the waterproofing membrane (i.e., drainage panel, retention mat, root barrier, and insulation board), but not the membrane itself. Use wind uplift design pressures and a minimum safety factor of 0.85 (or as defined by the designer) for wind uplift calculations where growth media is used to ballast.
- Where vegetated, pre-cultivated mats are installed, anchor them until the mat's root growth has achieved sufficient attachment into the growth media to adequately resist wind action (at least one full growing season). Vegetated mats should be properly anchored or ballasted against wind forces based on a safety factor of 1.0 (or as indicated by the designer).
- FM document also recommends the inclusion of Non-Vegetated Border Zones in the areas affected by larger wind pressure (edges and corners). These areas should be free of vegetation and covered by stone ballast or concrete pavers, and have a minimum dimension of 2.6 m. It also limits the installation of green roofs in geographical areas where the wind speed (3 seconds gust) does not exceed 45 m/s, applicable to all building heights as per FM document Data Sheet 1-35 Section 2.2.2."

Irrigation

The irrigation topic is explained in §**5.3 Irrigation (p. 12)**:

"OBC Requirements:

Division B 7.6.2.2

(1) Every potable water system that supplies a fixture or tank that is not subject to pressures above atmospheric shall be protected against back-siphonage by a back flow preventer.

(2) Where a potable water supply is connected to a boiler, tank, cooling jacket, lawn sprinkler system or other device where a non-potable fluid may be under pressure that is above atmospheric or the water outlet may be submerged in the non-potable fluid, the water supply shall be protected against back flow by a back flow preventer.

(3) Where a hose bibb is installed outside a building, inside a garage, or where there is an identifiable risk of contamination, the potable water system shall be protected against back flow by a back flow preventer.

City of Toronto By-law No.583-2009 § 492-9 M.

M. Irrigation.

Adequate measures shall be provided to permit irrigation necessary to initiate and sustain the vegetation during the service life of the green roof."

Maintenance/Cost estimation

The maintenance topic is discussed in §**6.2 Maintenance Plan (pp. 12–13)**:

"City of Toronto By-law No.583-2009 § 492-3.

Maintenance of Green Roofs and Health of Vegetation.

Every Green Roof required to be constructed pursuant to this Chapter shall be maintained in accordance with the maintenance plan required in the Toronto Green Roof Construction Standard.

City of Toronto By-law No.583-2009 § 492-9 N.

N. Maintenance Plan

(1) The Applicant shall develop a maintenance plan for the Green Roof as per CSA-S478-95 "Guideline on Durability in Buildings" which shall define programs of routine maintenance and inspection sufficient to ensure that the Green Roof components perform their required functions for the duration of their design service lives.

(2) The maintenance plan shall address the requirements of the specified growth media and vegetation for vegetation survival.

(*Continued*)

(3) The maintenance plan shall address re-planting, in the event that re-planting should become necessary, and assure that complete coverage at canopy level is achieved within three growing seasons and maintained for the service life of the Green Roof.

(4) The maintenance plan shall be submitted with the application for a permit for a Green Roof.

Maintenance Plans are focused primarily on the needs of the green roof during the first two to three years, or ongoing maintenance beyond three (approximately) years. Each type of maintenance plan should address the following matters:

(a) Protection measures for vegetated areas from impact of foot traffic (i.e., minimizing potential for damage);

(b) The amount, time and control of Irrigation;

(c) The amount, type and time of fertilizer application;

(d) Weed Management, including how often, which weeds to remove;

(e) Replanting (if required);

(f) Periodic inspection of drains, edging, roof penetrations, flashings, non-planted areas. This may also be part of the roofing membrane maintenance program;

(g) Troubleshooting; and

(h) Assignment of responsibility for removal of, care for, and reinstallation of growing medium and vegetation in the event of a leak."

CHAPTER 11

Research Reports

Research Report (1) Department for Communities and Local Government-London, 2013, Fire Performance of Green Roofs and Walls

This document provides comments on fire prevention strategies for green roofs and façades and references fire tests performed on vegetative systems, but it is not considered a standard, design guide, or research report. It reviews existing guides available regarding the fire performance of green roofs and façades and illustrates the results of the tests.

The document attempts to provide answers for the following questions regarding green roofs and façades:

(1) Is there a need to introduce additional fire prevention requirements for green roofs and façades?

(2) Is there a reasonable probability that green roofs and façades could ignite and fire could spread?

(3) Are there any inherent gaps in the functional requirements of case of green roofs and façades in the existing UK Building Regulations?

Applicability

(1) Fire performance and prevention strategies for green roofs and façades.

Structure

The document is divided into four chapters and one appendix:

1. Description of types of green roofs (extensive, semiintensive, and intensive) and green walls (climbing plants, hydroponic green walls, and modular green walls) and introduces features of green systems (focusing on fire design);
2. Several standards and documents regarding fire prevention and wind design are analyzed, including FLL, Factory Mutual Global Property Loss Prevention Data Sheets 1-35 – Green Roof Systems (2007), ANSI/SPRI VF-1 External Fire Design Standard for Vegetative Roofs, GRO Green Roof Code (2011), and Bridging the Gap – Fire Safety and Green Building;

Green Roofs, Facades, and Vegetative Systems
ISBN 978-0-12-817694-8
https://doi.org/10.1016/B978-0-12-817694-8.00011-X

3. Reference to the UK Building Regulations for the factors constituting fire risk. (i.e., How hard it is to ignite organic material? What is the likelihood and speed that fire spreads through organic material? What is the effect that spread of fire will have on the building?);
4. Requirements of UK Building Regulations that can be applied to green roofs and underlines the gaps in the regulations.
   Appendix A
   Summary of some fire test results on green roofs and façades

Characteristics

The document raises good questions about fire risk and raises a valid method for investigating the inherent gaps in building regulations for fire prevention of green roofs and façades.

The referenced documents/standards are not quoted. Also, the details of the fire tests on green roofs and façades are not exhaustively described (i.e., methods, instruments, input data, output data, etc., are not provided).

About
Fire risk/Fire safety

In §**1.3 Features of green roofs and walls (p. 9)**, the purpose of the document is explained: "It has been suggested that green roofs and walls may constitute a fire hazard [no references].

In relation to fire performance, the general consensus is that as long as the green roof or wall is kept moist (which is the normal case to keep the roof alive) it is likely to be very resistant to ignition. However, concern has been raised that if the green roof or wall dries out (such as might happen in a drought if no irrigation is provided), then they might present more of a fire risk. There has historically been a limited amount of fire testing of green roofs (carried out in Germany), but this does not necessarily address all the potential concerns. The other factor is that there is no significant fire testing of green wall systems. For additional requirements to be introduced for green roofs or walls, it would need to be demonstrated that there is a reasonable probability that they could ignite and that the fire could spread in such a way that it would cause a breach of the functional requirements of the Building Regulations."

§**2.3 Prevention of fire (pp. 11–13)** lists the design guidelines from the four main international standards: FLL, ANSI/SPRI VF-1, FM Datasheets 1-35, GRO: "The guidance reviewed generally states that green roofs should be designed to provide the necessary resistance to the external spread of fire by the following measures:

(1) Increasing the non-combustible content of the growing medium;

(2) Decreasing the organic content of the growing medium;
(3) Preventing the system from drying out.

Extensive roofs are not generally irrigated therefore the fire risk is mitigated by the specification of the build up and fire breaks, and by reducing the organic content."

Standard	Layering	Firebreaks	Fire hydrants
FLL	Considers a well-maintained and irrigated green roof as a hard roof.	**(1)** Firebreaks wide =500 mm, installed all around openings of the roof and vertical elements; **(2)** Firebreaks = 800 mm installed at sills; **(3)** Firebreaks wide 1m every 40m of roof; **(4)** End walls, firewalls, and separating walls should be of height 300 mm above the substrate every 40m. If end walls, firewalls, and separating walls do not extend above the substrate, then noncombustible top piece or a concrete slab high, 300 mm, is acceptable.	
ANSI/ SPRI VF-1	• Recommended succulent plants and grass-based systems; • Substrate inorganic content >80%.	**(1)** The fire breaks should be class A (as for ASTM E108 and UL 790) rated roofing system for a minimum 1.8 m wide continuous border. **(2)** Green roof sections are limited to 1450 m^2 with a maximum wide of 39m and must be divided by firestop walls of noncombustible construction extended above the roof surface by minimum 914 mm and a fire break border 1.8 m wide.	The access of one or more fire hydrants should be provided.

Continued

Standard	Layering	Firebreaks	Fire hydrants
FM datasheet 1-35	**(1)** Vegetation height <0.9 m; **(2)** (Avoided grasses and mosses plants.	**(1)** Recommends the provision of the continuous 500 mm width border around all roof top equipment; **(2)** Perimeter parapet walls should be provided for all green roof systems, which should extend a minimum of 150 mm in elevation above the top of the growth media. Where roof elevation is above 46 m, the perimeter parapet walls should be a minimum of 760 mm in height; where parapet walls of these height are provided, the adjacent border zone (free of vegetation and growing media) may be reduced to a minimum of 0.9 m.	
GRO	**(1)** Substrate depth >30 mm; **(2)** Substrate organic content <20%; **(3)** Recommended succulent plants.	Same as FLL.	

In §3 **Risks of green roofs/walls in relation to Building Regulations requirements (pp. 16–17),** the functional requirements inherent to fire prevention are listed as follows:

(1) B1—Means of warning and escape

(2) B2—Internal fire spread (linings)
(3) B3—Internal fire spread (structure)
(4) B4—External fire spread
(5) B5—Access and facilities for the fire service

It is specified that: "When considering the risk associated with green roofs and wall their use could potentially affect B2, B3, B4, B5. At this time there do not appear to be any significant issues identified with green roofs/walls in relation to means of escape per se.

When considering the internal fire spread (linings) there may be potential for flames to spread through a green roof deck into a building and vice versa.

When considering the internal fire spread (structure) there may be potential for fire spread over a compartment wall or across a roof into a compartment (stepped façade). There may also be potential for fire spread across a wall façade which could compromise compartmentation, or a detachment of the wall façade may occur.

When considering the external fire spread there may be a risk of ignition of, or ignition by, an adjacent building by radiant heat and/or burning embers.

When considering access and facilities for the fire and rescue service there may be a risk that fire spread across a wall façade could be extensive and difficult to fight."

In §4 **Building regulations requirements (pp. 18–30)**, the compliance of green roofs and green walls to functional requirements inherent to fire prevention (B2, B3, B4, B5 above mentioned) is analyzed for each requirement. The table below shows the main contents and quotes the tests described in Appendix A:

		Compliance with building regulation requirements		
	Building regulations requirements	**§4.5 Recommendations for green roofs**	**§4.5 Recommendations for green walls**	**Test in case of breach in the building regulations**
B2	To inhibit the spread of fire within the building, the internal linings shall: (1) adequately resist the spread of flame over their surfaces; and (2) Have, if ignited, a rate of heat release or a rate of fire growth, which is reasonable in the circumstances.		If the use of a living wall system is proposed, then test evidence should be provided to demonstrate that it has the appropriate classification for the relevant location in the building. It should be noted however that parts of walls may be of a poorer performance specified in Table 10 of Approved Document B as long as they are not poorer than Class 3 (National class) or Class D-s3, d2 (European class). These areas are limited to less than half the floor area of the room, and cannot be more than $20m^2$ in residential accommodation or $60m^2$ in non-residential accommodation.	

B3	**(1)** A wall common to two or more buildings shall be designed and constructed so that it adequately resists the spread of fire between those buildings; **(2)** Where reasonably necessary to inhibit the spread of fire within a building, measures shall be taken, to an extent appropriate to the size and intended use of the building, comprising either or both of the following; **(3)** Sub-division of the building with fire resisting construction;	**(1)** In order to comply with this requirement there should be adequate provisions to prevent a fire in the top floor of a building from breaking through the ceiling into the green roof, then spreading horizontally over a compartment wall and back down into an adjacent compartment; **(2)** Where compartment walls meet the underside of the roof deck, they should be fire-stopped and a zone of 1500 mm wide, on either side of the wall should have a covering designation AA, AB or AC on a material	**(1)** In order to comply with this requirement there should be adequate provisions in place so that a fire is prevented from spreading up the wall which could cause the fire to spread from one floor to another across a compartment floor; **(2)** Where green wall systems are fixed to a wall which contains compartment floors the guidance in Diagram 33 of Approved Document B Fire Safety (of the Building Regulation in UK) should be followed and fire stopping should be provided between the compartment floor	**(1)** Appendix A Scenario 1 **(2)** Appendix A Scenario 2

Continued

Building regulations requirements	Compliance with building regulation requirements: §4.5 Recommendations for green roofs	Compliance with building regulation requirements: §4.5 Recommendations for green walls	Test in case of breach in the building regulations
(4) Installation of suitable automatic fire suppression systems.	of limited combustibility; **(3)** The use of fire breaks around all openings and vertical elements are strongly recommended for all types of green roofs (the fire break should consist of paving slabs or non-vegetated strips of pebbles with a depth of 75 mm and diameter of 20 and 50 mm for a width of 500 mm; **(4)** Whilst fire spread to a green roof via penetrations such as roof lights, pipes or vents in the roof is not in itself a breach	and the external wall. The fire stopping should provide the same fire resistance as that required for the compartment floor; **(3)** Where a green wall system is fixed to an external wall and a cavity is formed then the size of the cavity should be limited using horizontal and vertical cavity barriers according to the guidance given in Table 13 of Approved Document B.	

		of requirements B1 to B5, it is recommended that fire breaks are provided around such penetrations so that basic maintenance procedures can be carried out; **(5)** Testing carried out on the growing media using the cone calorimeter showed that for substrates using leaf mold as the organic material ignition occurred at concentrations greater than 50%. The sample containing 100% leaf mold when completely dried was very flammable, which could be seen to represent the fire spread and growth		

Continued

Building regulations requirements	Compliance with building regulation requirements: §4.5 Recommendations for green roofs	Compliance with building regulation requirements: §4.5 Recommendations for green walls	Test in case of breach in the building regulations
	conditions in a forest fire. This however is considered an extreme situation as it is highly unlikely, given the climate in the United Kingdom, that a green roof which has not been maintained and irrigated for a significant amount of time would have dried out to such an extent. Furthermore, it is also highly unlikely that the growing media would ever consist entirely of leaf mold; **(6)** Nevertheless, it is recommended that		

		the recommendations of the existing guidance are followed and that fire breaks are provided in 1m strips every 40m across extensive green roofs.		
B4	**(1)** The external walls of the building shall adequately resist the spread of fire over the walls and from one building to another, having regard to the height, use and position of the building; **(2)** The roof of the building shall adequately resist the spread of fire over the roof and from one building to another, having regard to the	**(1)** In order to comply with this requirement there should be adequate provisions in place so that a fire in an adjacent building which causes burning brands and/or radiation onto an adjacent building with a green roof will not result in fire spread through the green roof and into the building; **(2)** It is considered that if a fire in a green roof	See Fig. 2. Approved Document B Diagram 40 – provision for external surfaces or walls. • The provisions for the external walls depend on the height of the building, the purpose group and the distance to the relevant boundary; • For all buildings (except assembly and recreation with more than one storey) where the height of the building is less than 18m and the	**(1)** Appendix A Scenario 3: the test also measured the temperatures at different depths in the growing media to assess whether the temperatures would be high enough to cause ignition of the layers below; **(2)** The test determined whether there was any fire penetration through the growing media and also measured the

Continued

Building regulations requirements	Compliance with building regulation requirements: §4.5 Recommendations for green roofs	Compliance with building regulation requirements: §4.5 Recommendations for green walls	Test in case of breach in the building regulations
use and position of the building.	only involved the vegetation, for example a bush in an intensive roof or local charring of vegetation, then the risk was negligible as it would be no different from similar vegetation in a normal garden. However, if the fire penetrated the growing layer and broke through the roof into the storey below then this would be a breach of Regulation B4 and **(3)** Large scale tests using the European standard DD CEN/TS 1187	distance to the relevant boundary exceeds 1m there are no provisions (Figure b in Diagram 40). Given that there are no provisions green walls are permitted in such buildings as they would present no greater fire risk than a building façade lined with a combustible material; • Approved Document B also states that in a building with a storey 18m or more above ground level any material used in the external wall construction should be of limited	temperatures at different depths in the growing media to assess whether the temperatures would be high enough to cause ignition of the layers below.

	were carried out on the growing media used in green roof systems with different concentrations of organic matter. Further details of the test and results are given in Appendix A.3. The test determined whether there was any fire penetration through the growing media and also measured the temperatures at different depths in the growing media to assess whether the temperatures would be high enough to cause ignition of the layers below; **(4)** The tests on the growing media showed that for the	combustibility. Diagram 40 outlines two other provisions for external surfaces (Figures c and e). In assembly and recreation buildings with more than one storey up to a height of 10m, and up to a height of 18m in buildings with a height greater than 18m the external surfaces should have an index (I) not more than 20 (national class) or class C-s3, d2 (European class) as a minimum; • For buildings with a height greater than 18m where the boundary distance is less than 1m and at heights above 18m in tall buildings the external surfaces should be class	

Continued

Building regulations requirements	Compliance with building regulation requirements		Test in case of breach in the building regulations
	§4.5 Recommendations for green roofs	§4.5 Recommendations for green walls	
	most onerous (albeit unlikely) case of a growing medium consisting of 100% leaf mold the temperature at the lowest point in the substrate was not sufficient to ignite the materials used in the layers between the growing media and the roof deck. Consequently, a fire on the green roof would not penetrate the layers beneath the growing media and it is highly unlikely that it would penetrate the structural deck below;	0 (national class) or class B-s3, d2 (European class) as a minimum.	

		(5) In order for green roofs to comply with requirement B4 it is recommended that for all types of green roof the depth of the growing layer should be a minimum of 80 mm and the organic content should not exceed 50%.		
B5	The building shall be designed and constructed so as to provide reasonable facilities to assist fire fighters in the protection of life.			

§**Appendix A – Experimental Work (pp. 31–36)** describes four scenarios representing breaches for the requirements previously listed: B2—internal fire spread (linings); B3—internal fire spread (structure); B4—external fire spread; and B5—access and facilities for the fire service. It summarizes the experimental tests done with a cone calorimetry on both green roof and wall systems, and large-scale tests on substrates for green roofs (using the DD CEN/TS 1187).

§A1 Scenarios considered in Appendix A – Experimental Work **(p. 31)** lists the four scenarios:

(1) Scenario 1 considers a fire on the top floor of a building with a green roof with a breach of the requirement B3.

(2) Scenario 2 considers fire spread up a green wall causing radiation and consequently fire spread to adjacent buildings. This would be a breach of the requirement B4.

(3) Scenario 3 considers a fire in an adjacent building where burning brands or radiation on to an adjacent green roof causes fire spread with a green roof. This would be a breach of the requirement B4.

(4) Scenario 4 considers the fire spread from within a building with a green roof via penetration such as roof lights, pipes, or vents in the roof. This would be a breach of the requirement B5.

§**A2 Cone calorimetry (p. 32)** summarizes some tests carried out on green roofs and façades. The fire tests investigate the fire performance of the substrate varying of organic (standard mix, leaf mold, and recycled compost) and inorganic components (crushed bricks and pumice).

The tests are not extensively described.

Test description	Results
(1) Conditioning room and oven at 40°C **(2)** Cone calorimetry	**(1)** No ignition occurred for substrate using standard organic mix even where completely dried; **(2)** With a completely dried 100% organic mix where ignition occurred, the flame spread could not be sustained because the peak heat release rate was 55Kw/m^2; **(3)** For substrate using leaf mold as organic material, ignition occurred at concentration greater than 50%. The sample containing 100% leaf mold, when completely dried, was very flammable.

Then, the test on living walls is described, as for green roofs, to investigate the fire performance of the substrates:

Test description	Results
(1) Conditioning room and oven at 40°C **(2)** Cone calorimeter at 50 kW/m^2 **(3)** Growing substrate consisting of medium textured peat and coir with added wood fiber, perlite, lime base nutrients, and CRF fertilizer with added wetter and insect control	**(1)** No ignition was observed in any of the tests. No signs of possible ignition, therefore the reactions were not considered no-flaming. It is therefore unlikely that the substrate could ignite or contributed to flames spread even when completely dried out.

§**A3 Large-scale testing (pp. 32–33)** summarizes new tests (using DD CEN/TS 1187) with the aim to determine the fire impact on the surface of the green roofs:

Test description	Results
• Angle of 20°; • Sample 80 mm thick (further testing at 120 and 150 mm depths); • Observation after 60 min.	• No fire penetration on the base of the samples; • Temperature measured at the lowest point of the substrate was generally less than 100°C; • Temperature measured at the lowest point of the substrate at 218°C in the case of 100% dried out leaf mold (the polymers normally used in green roof for the components below the substrate have piloted ignition temperature greater than 218°C).

Five other tests on shelf green wall systems (using BS EN 12,823) are briefly discussed (i.e., Biotecture with Hydroponic system, ANS with filled HDPE modules, Optigreen with aluminum mesh cassette; Cityroof with porous plastic irrigated board; Minigarden with Stacked HDPE planters).

All the systems failed the tests because the substrate used in living walls could not be ignited, even when completely dried out.

The conclusions of the tests are as follows:

- In general, the substrate used for both green roofs and walls cannot be ignited and fire spread does not occur;

- The temperatures obtained at the lowest point of the substrate layers are not high enough to result in ignition of material used below the substrate;
- Ignition only occurs when the substrate consists of 100% leaf mold completely dried out (conditions highly unlikely);
- Substrates used in living walls cannot be ignited, even when completely dried out.

Structural safety/Wind design

The wind topic is not deeply discussed.

In §**2.2 Wind design (p. 11)** some other documents are mentioned: "Each document recognizes that wind design is the first governing criteria when determining if a green roof should be considered for a building. Fire safety design should follow subsequent to this approval."

The mentioned recommendations are the following:

From	Contents
FM Data Sheet 1-35	Green roof systems should only be installed in geographical locations where the basic wind speed (3 s gust), as determined from FM Data Sheet 1–28 design wind loads, does not exceed 100 mph (45 m/s). Figure 18 of FM Data Sheet 1–28 gives the basic wind speeds in miles per hour for Europe, and for the majority of countries (with the exception of the far west of Ireland and north and west Scotland) the wind speed is less than 45 m/s.
European guidance	Makes less reference to the wind speed but does state that green roof designs should comply with all relevant structural design criteria. In addition, where green roof elements are being used as ballast to prevent items that are not mechanically fixed from wind uplift, sufficient weight must be incorporated into the green roof build up and dry weights must be used to calculate the weight of the green roof system. Traditional "Nordic" style pitched green roofs frequently occur in locations which have high winds however the green roofs are used to weigh down the waterproofing layer. This practice is contradictory to the advice in the FM Data Sheet however the European guidance allows for this."

Irrigation

§**1.1 Green roof (pp. 5–6)** explains: "An extensive roof does not normally require irrigation and is low in nutrients.

For a semi-intensive roofs: the GRO Green roof code guidance states that irrigation and maintenance requirements are dependent upon the plant species installed, whereas the Factory Mutual Datasheet 1-35 recommends a permanent irrigation system for this type of roof.

For an intensive roof, the Factory Mutual Datasheet 1-35 recommends a permanent irrigation system."

§**1.2 Green walls (pp. 7–8)** explains: "Climbing plants system are usually irrigated but can survive without irrigation if rooted into the ground.

In Hydroponic green walls, plants grow without substrate or soil and rely of irrigation and nutrients added to irrigation water. Irrigation water is usually delivered to the top of each module via irrigation lines."

Maintenance/Cost estimation

The maintenance topic is not deeply discussed.

§**4.6.4 Installation and maintenance (pp. 29–30)** makes some specifications about façades: "Typical green walls are not designed to be "no maintenance" and irrigation is generally required otherwise the plants will not thrive however some systems require less water than others. The living wall systems should be regularly maintained to ensure that the irrigation system is working. Maintenance contracts can be arranged, generally on an annual basis however contracts up to 10 years have been known."

Research Report (2) National Association of State Fire Marshals: Tidwell, Jim and Jack J.Murphy, 2010, Bridging the Gap: Fire Safety and Green Buildings. A Fire and Building Safety Guide to Green Construction

The research analyzes green buildings (building envelopes, design attributes, building systems, and alternative power sources) and provides a series of checklists that could be used a "starting point" for fire safety and green construction and as a tool to begin a serious dialogue between the fire services and those involved in green construction.

Applicability

- The research examines several green roof elements and construction systems.

Structure

The research is divided into five chapters and six appendixes:

1. Introduction: the current status of the Green Movement in the United States and the world; the importance of energy conservation and green certification; green building and green component concepts; fire safety considerations that are lacking from codes, standards, and rating systems; the importance of fire safety measures for green buildings, to increase public safety and reduce the risk to firefighters and the impact that a fire event has on the environment (see Appendix F: controlled burn air emission tables)
2. Site selection and use: factors to consider in selecting the function and site for green construction (i.e., firefighter access, building orientation, and surrounding landscape)
3. Analysis of design considerations that can contribute to sustainability: building shell components (insulation, vegetative roof system, high performance glazing); design attributes (ventilation controlled fire, large open space and automatically controlled windows for energy); skylights and solar tubes; water conservation; structural concerns; miscellaneous components (vestibules, awnings, and recycling practices)
4. Building systems and alternative power sources (i.e., photovoltaic solar power system, wind turbine power systems, hydrogen fuel cell power system, battery storage system, nuclear-generated power, HVAC system, and high-volume/low-speed [HVLS] fans)
5. Conclusions

Appendix:

A Matrices identifying green building components and their respective concerns

B Glossary of terms (incl. references to vegetative roofs)

C Name and description of the leading national organizations for sustainable construction

D List of some case studies

E List of the quoted references

F Table about controlled burn air emissions from FM Global

Characteristics

The fire concerns for each green element/construction system are clearly listed. Multiple concerns are addressed for green roof systems: roof loads including wind uplift pressures and wind loads, roof drainage (closely linked to roof load issue), roof slope, roof access, load-bearing structures, parapets, and fire exposure (with reference to FM Global).

About

Fire risk/Fire safety

In §**B Vegetative roof (pp. 14−18)**, of **Chapter 3: Building envelope and design attributes**, the three types of green roofs are described: extensive, intensive and simple intensive.

Issues regarding access are highlighted: "Roof access provisions should be made for firefighters. A clear space around the roof perimeter should be wide enough to accommodate firefighters and their equipment. In addition, pathways should be provided so that firefighters can access rooftop equipment, skylights, etc., so that vertical ventilation operation can be undertaken. FM Global suggests that green roofs should be divided into section no longer than 15,625 square feet, with no dimension greater than 125 feet. Section should be separated by non combustible walkways wide enough to accommodate a firefighter's equipment."

Then, issues with fire exposure are highlighted: "while there are no current rating of green roof FM suggests that green roofs should be considered similar to traditional roofs with regard to exposure from an interior fire – if the roof is metal deck, it will be considered a Class I or II; if it is concrete deck, it will be considered non-combustible. When considering exterior fire exposure, a careful evaluation is necessary. It is this hazard that should be considered when selecting plants and other materials that will be on the roof. Plants with high moisture content, such as low-growing succulents and similar plants, will enhance the roof's fire performance, assuming they are maintained. Plants with high level of volatile oils or resins should be avoided, especially in areas prone to wildland fires."

Appendix A: Checklist (pp. 43−45) provides a series of checklists in the form of matrices that identify green building components and relate each component to a series of potential concerns (vegetative roofs are mentioned).

Structural safety/Wind design

§**B Vegetative roof (pp. 14–18)** explains: "wind uplift pressures and wind loads should be included in the process of designing the structural system. Particular attention should be given to areas prone to high winds, such as those that occur during hurricanes. All calculations for wind events should be made in a dry condition and without a vegetation load. Any roofing materials or objects placed on the roof that might become flying debris should be avoided."

Irrigation

N/A.

Maintenance/Cost estimation

§**B Vegetative roof (pp. 14–18)** describes the level of maintenance for the three types of green roofs: intensive (extensive maintenance, including mowing, irrigation, and trimming), extensive (low to none; plants are self-sustaining after two or three growing season), and simple-intensive (this hybrid type may include plants that require greater maintenance and irrigation).

Research Report (3) FPRF-Fire Protection Research Foundation, 2012, Fire Safety Challenges of Green Buildings: Final Report

This is the final research report of a study identifying gaps and specific research needs in understanding and addressing fire risk and hazards with green building design. The results of the study include: a list of fire incidents involving green elements or green buildings (no green roofs/façades are mentioned); a list of fire hazards and risk factors associated with the identified green building elements; strategies to mitigate fire risk for green building elements; and an analysis of the IgCC and green rating schemes (i.e., LEED, BREEAM, GREENMARK), identifying fire safety measures that are not considered.

Applicability

- Analysis of fire protection concerns for green roofs, green façades, and several types of "green elements" and "attributes" of green construction.

Structure

The standard is divided into seven sections and seven appendixes:

1 Background, introduction, and project objectives (i.e., identifying green building elements that are fire hazards and to share best practices for fire risk mitigation)

2 Summary of the research objectives, a detailed description of the sources, the research approach/methods, and listing fire incident reports, news reports, fire investigation reports, and reports related to fires in green buildings and green elements

3 Attributes of green buildings and elements, which are summarized in Table 4 (where green roofs and façades are cited

4 Attributes of green buildings or sites which could impact fire, life safety, building or fire service performance, and objectives (typically addressed by building/fire codes and are identified in the literature review), which are summarized in Table 5

5 The main chapter of the research, which summarizes the hazard and risk assessment into three main results:

- **(1)** A matrix for evaluating the potential fire hazard of green elements/attributes
- **(2)** A table with a list of potential mitigation strategies associated to the green elements/attributes
- **(3)** Schematic representations of fire hazards, which could serve as a quick overview guide or fire performance ranking for a building

6 Summary and conclusions

7 References

Appendix:

A Informational resource list of the sources, adding to the contents introduced in §**2**

B A selection of potential fire concerns with green building elements and mitigation strategies, adding to the contents introduced in §2.1 and §2.2

C International surveys and responses, which include inquiries and answers from experts of international fire institutions and building regulatory agencies, adding to the contents introduced in §**2.3**

D Detailed matrices of the potential fire hazards of many green elements (i.e., structural materials, exterior materials, interior materials, building systems, etc.), adding to the contents introduced in §**5.1 (Fig. 1)**

E An illustrated matrix identifying the relative hazard ranking of green elements, adding to the contents introduced in §**5.1 (Fig. 2)**

F Tables of fire hazards with green building elements and potential mitigation strategies, adding to the contents introduced in §**5.2**

G Tables assessing LEED, BREEAM, GREENMARK, and IgCC for fire safety objectives, including schematic drawings of the fire interaction and adding to the contents introduced in §5.3 (Fig. 7)

Characteristics

As a complete research study, an in-depth methodology, summary, and various graphics are provided, which make the research a good tool for engineers, designers, insurers, authorities, or others when reviewing site plans, building designs, renovation designs, or buildings to assist in hazard or risk analysis, and for developing fire mitigation strategies. Also, four green building codes are analyzed and case studies are identified, which could help highlight fire protection issues for green living technologies that may not have been considered in green building codes or past projects.

This research report specifies that there are limited incidents of fires in documented green buildings, so there is not sufficient data to provide a quantified risk assessment. Due to the lack of data for actual fire incidents on green building elements, a ranking of their risks/hazards was the approach used in this research.

About

Fire risk/Fire safety

In this research, green roofs and façades are mentioned as elements that are have a moderate to low level of concern for fire hazards.

Appendix F Tabular Representation of Fire Hazards with Green Building Elements **(pp. 69–71)** contains mitigation instructions for green roofs and facades:

Material/ System/ Attribute	Hazard	Concern level	Potential mitigation strategies
Vegetative roof system **(p. 69)**	Can contribute to fire loads, spread of fire, impact FF operations, impacts smoke and heat venting, contribute to stability issue.	Moderate	Manage fire risk of vegetation. Assure use of fire tested components. Provide adequate area for FD access, smoke/heat venting, and other operations. Approved/ listed materials.

Material/ System/ Attribute	Hazard	Concern level	Potential mitigation strategies
Exterior vegetative covering **(p. 69)**	Can impact FF access and present WUI issue.	Low	Limit volume.
Interior vegetation **(p. 70)**	Can contribute to fuel load. Can impact FF operations.	Low	Sprinklers.
Vegetation for shading **(p. 71)**	May present fire apparatus access challenges.	Low	Appropriate emergency response planning, including vehicle access.

Appendix B: Representative Fire Issues and Mitigation Approaches (pp. 42–53) declares that green roofs may include flammable materials, flammable vegetation with a risk of fire spread, and firefighter access issues.

The authors quote two reference standards:

- Approval Standard for Vegetative Roof Systems—FM Approvals—contains general information, general requirements, performance requirements, and operations requirements
- ANSI/SPRI VF-1 External Fire Design Standard for Vegetative Roofs

Appendix E: Illustration of Relative Hazard Ranking with Detailed Matrix (pp. 64–67) provides the hazard ranking of vegetative roof systems, exterior vegetative coverings, interior vegetation, and vegetation for shading:

Stuctural Materials and Systems	Poses potential ignition hazard	Poses potential shock hazard	Potential explosion hazard	Poses potential toxicity hazard	Readily ignitable	Burns readily once ignited	Contributes more fuel / increased HRR	Material affects burning characteristics	Fast(er) fire growth rate	Significant smoke production / hazard	Potential for shorter time to failure	Failure affects burning characteristics	Failure presents stability concern	Failure presents smoke spread concern	Failure presents flame spread concern	Material presents flame spread concern	May impact smoke / heat venting	May impact occupant evacuation	May impact FF water availability	May impact suppression effectiveness	May impact fire apparatus access	May impact FF access and operations	May impact containment of runoff	Relative risk level
Vegetative roof systems					Moderate	Moderate	Moderate									High						High		Moderate
- Insulating material							Moderate									Moderate								Moderate
- Thickness																						High		High
- Type of vegetation																Moderate								Moderate
Exterior vegetative coating						Moderate	Moderate									Moderate		High		Moderate	Moderate	Moderate		Moderate
Interior vegetation							Moderate											Moderate						
Vegetation for shading																		Moderate			Moderate			Moderate

Key:

Low: presents a low risk when unmitigated or is not applicable to the listed attributes

Moderate: presents a moderate risk when unmitigated

High: presents high risk when unmitigated

Relative fire risk/Hazard level of structural materials and systems **(p. 65)**. Data relating vegetative systems.

Appendix G: Assessment of LEED, BREEAM, GREENMARK and the IgCC for Fire Safety Objectives (starting on p. 72) reviews four green rating schemes/codes (i.e., BREEAM, GREENMARK, LEED, IgCC) and identifies fire safety concerns for each.

"For the review, each rating system/code that was reviewed was converted into the format below [see the extracts in the next pages], identifying Category/Chapter in the document, Section/Assessment issue being addressed, Aim of the scheme (where stated), Credits/Scores (where provided/used in the scheme), specific Requirements/Criteria required to be achieved, Procedures to be followed in achieving the targeted performance, Primary or Secondary fire hazards that were identified as part of the review, and a Summary of the potential fire impact represented by a schematic illustration."**(p. 26)**

The extracts below are the assessment tables for the green rating schemes/codes that relate to green roofs, green façades, and vegetation surrounding a building:

BREEAM:

Categories / Chapter	Section / Assessment Issues	Aims	Credits / Scores	Requirements / Criteria	Procedures	Fire Hazard Primary (P) & Secondary (S)	Summary
Health and Wellbeing (Hea)	Hea 01- Visual comfort	To ensure daylight, artificial lighting and occupant controls are considered at the design stage to ensure best practice visual performance and comfort for building occupants	BTD		Daylighting design needs a less compact Building Geometry	(P) May influence fire and smoke exterior spread. More transparent surfaces on facades could influence O2 presence (S) May affect evacuation routes and fire services intervention	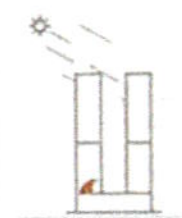
					View out and glare requirements	(P) Glare control systems may influence exterior spread and fire services intervention. In the case of green elements, combustible material added	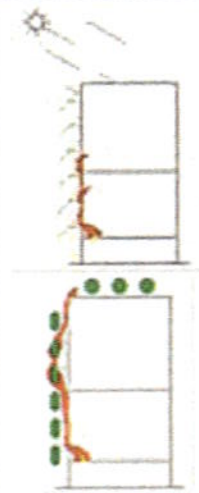
	Haa 02- Indoor air quality	To recognise and encourage a healthy internal environment through the specification and installation of appropriate ventilation, equipment and finish	BTD		Minimising sources of indoor air pollution	(P) Air conditions may influence O2 presence and spread	
					Potential for natural ventilation	(P) Air movement within the building may affect fire and smoke spread	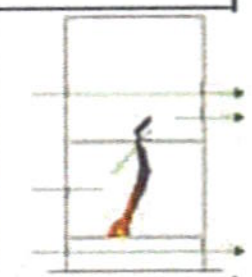

Categories / Chapter	Section / Assessment Issues	Aims	Credits / Scores	Requirements / Criteria	Procedures	Fire Hazard Primary (P) & Secondary (S)	Summary
Energy (Ene)	Ene 01- Reduction of CO2 emissions	To recognise and encourage buildings designed to minimise operational energy demand, consumption and CO2 emissions	15		Energy Demand can be minimised with: • Insulation • Renewal energies • Facade systems or elements • Sustainable and recycled materials	P) The use of better insulation may increase interior temperature in a fire event (P) The use of renewal energies may affect fire and smoke spread (P) The use of façade systems or elements to regulate thermal conditions and sun radiation may influence fire and smoke spread (P) Encouraging the use of sustainable and recycled materials may be contradictory with diminishing combustion and/or toxicity (S) May influence evacuation or fire services intervention	
	Ene 02- Energy monitoring	To recognise and encourage the installation of energy sub-metering that facilitates the monitoring of operational energy consumption	BTD		Use a Building Energy Management System (BEMS) or separate accessible energy submeters	(P) Increasing the systems complexity may affect their response	
	Ene 03- External lighting	To recognise and encourage the specification of energy-efficient light fittings for external areas of the development	1			(P) May affect fire services intervention at night	

GREENMARK:

Categories / Chapter	Section / Assessment Issues	Aims	Credits / Scores	Requirements / Criteria	Procedures	Fire Hazard Primary (P) & Secondary (S)	Summary
	NRB 1-10 Energy Efficient Practices & Features Encourage the use of energy efficient practices and features which are innovative and/or have positive environmental impact		12	(a) Computation of energy consumption based on design load in the form of energy efficiency index (IEE) (b) Use of vertical greenery system on east and west façade to reduce heat gain (c) Use of energy efficient features Examples: · Heat recovery system · Sun pipes · Lifts with gearless drive · Re-generative lift · Light shelves · Photocell sensors to maximize the use of daylighting · Heat pumps etc.	Adding vertical greenery systems	(P) The use of greenery systems may influence fire initiation and fire exterior spread (S) May disturb fire brigade intervention	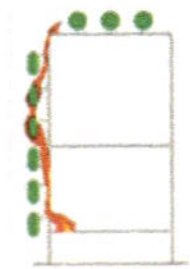
	NRB 1-11 Renewable Energy Encourage the application of renewable energy sources in buildings		20		· Use of solar photovoltaic systems · Use of solar water heating equipment · Use of wind energy systems	(P) The use of these systems may influence exterior spread (S) May disturb fire brigade intervention (i.e. access to the building or smoke venting)	
Water Efficiency (17 p. max.)	NRB 2-1 Water Efficiency Fittings Encourage the use of water efficient fittings covered under the Water Efficiency Labeling Scheme (WELS)		10	(a) Basin taps and mixers (b) Flushing cistern (c) Shower taps, mixers or showerheads (d) Sink/Bib taps and mixers (e) Urinals and urinal flush valve		(P) May affect the election of fire protection systems according to water use	

Categories / Chapter	Section / Assessment Issues	Aims	Credits / Scores	Requirements / Criteria	Procedures	Fire Hazard Primary (P) & Secondary (S)	Summary
	NRB 2-2 Water Usage and Leak Detection Promote the use of sub-metering and leak detection system for better control and monitoring		2	(a) Provision of private meters to monitor the major water usage such as irrigation, cooling tower and tenant's usage (b) Linking all private meters to the Building Management System (BMS) for leak detection		(P) Extending Monitoring may influence water systems complexity and increase the possibility of failure	
	NRB 2-3 Irrigation system and Landscaping Provision of suitable systems that utilize rainwater or recycled water and use of plants that require minimal irrigation to reduce potable water consumption		3	(a) Use of non potable water for irrigation (b) Use of automatic water efficient irrigation system (c) Use of drought tolerant plants to require minimal irrigation		(P) The use of drought tolerant plants may influence fire initiation and fire spread	
	NRB 2-4 Water Consumption of Cooling Towers Reduce potable water use for cooling purpose		2	(a) Use of cooling tower water treatment which can achieve 7 or better cycles of concentration at acceptable water quality (b) Use of NEWater or on-site recycled water from approved sources			
Environmenta l Protection (42 p. max.)	NRB 3.1 Sustainable Construction Encourage recycling and the adoption of building designs, construction practices and materials that are environmentally friendly and sustainable		10	(a) Use of Sustainable and Recycled Materials (i) Green Cements (ii) Recycled Concrete Aggregates (b) Concrete Usage Index (CUI)		(P) Environmentally friendly designs, construction practices and materials may influence combustion, fire and smoke spread and toxicity	

Categories / Chapter	Section / Assessment Issues	Aims	Credits / Scores	Requirements / Criteria	Procedures	Fire Hazard Primary (P) & Secondary (S)	Summary
	RB 1-5 Ventilation Carparks Encourage the use of energy efficient design and control of ventilation systems in car parks		6	(a) Car parks designed with natural ventilation (b) CO sensors to control mechanical ventilation		(P) The use of natural ventilation in car parks may influence fire exterior spread	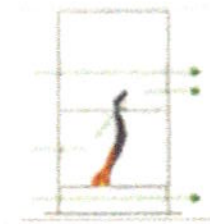
	RB 1-6 Lifts Encourage the use of lifts with energy efficient features		1				
	RB 1-7 Energy Efficient Features Encourage the use of energy efficient features		7	Examples: · Lifts with gearless drive · Re-generative lifts · Heat recovery devices · Cool paints · Gas water heaters · Calculation of EEI · Provision of vertical greenery system that helps to reduce heat gain	Adding vertical greenery systems	(P) The use of greenery systems may influence fire initiation and fire exterior spread (S) May disturb fire brigade intervention	
	RB 1-8 Renewable Energy Encourage the application of renewable energy sources such as solar energy in build ons		20		· Use of solar photovoltaic systems · Use of solar water heating equipment · Use of wind energy systems	(P) The use of these systems may influence exterior spread (S) May disturb fire brigade intervention (i.e. access to the building or smoke venting)	

Categories / Chapter	Section / Assessment Issues	Aims	Credits / Scores	Requirements / Criteria	Procedures	Fire Hazard Primary (P) & Secondary (S)	Summary
Water Efficiency (14 p. max.)	RB 2-1 Water Efficiency Fittings Encourage the use of water efficient fittings		10	(a) Basin taps and mixers (b) Flushing cistern (c) Shower taps, mixers or showerheads (d) Sink/Bib taps and mixers (e) All other water fittings		(P) May affect the election of fire protection systems according to water use	
	RB 2-2 Water Usage Monitoring Provide private meters to monitor water usage		1			(P) Extending Monitoring may influence water systems complexity and increase the possibility of failure	
	RB 2-3 Irrigation system and Landscaping Provision of suitable systems that utilize rainwater or recycled water for some uses		3	(a) use of non potable water for irrigation (b) Use of automatic water efficient irrigation system (c) Use of drought tolerant plants to require minimal irrigation		(P) The use of drought tolerant plants may influence fire initiation and fire spread	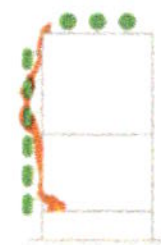
Environmental Protection (41 p. max.)	RB 3.1 Sustainable Construction Encourage recycling and the adoption of building designs, construction practices and materials that are environmentally friendly		10	(a) Use of Sustainable and Recycled Materials (b) Concrete Usage Index (CUI)		(P) Sustainable materials may influence combustion, fire and smoke spread and toxicity	
	RB 3-2 Sustainable Products Promote the use of environmental friendly products that are certified		8			P) Sustainable products may influence combustion, fire and smoke spread and toxicity	

IgCC:

Categories / Chapter	Section / Assessment Issues	Aims	Credits / Scores	Requirements / Criteria	Procedures	Fire Hazard Primary (P) & Secondary (S)	Summary
	408 Heat Island Mitigation						
	408.2 Site Hardscape			In climate zones 1 through 6, IECC, no less than 50% of the site hardscape provided with materials with adequate solar reflectance value, shading by structures, shading by trees or pervious and permeable pavement* (*) Permitted where the use does not interfere with fire and emergency apparatus or vehicle or personnel access and egress		(P) Could affect fire spread	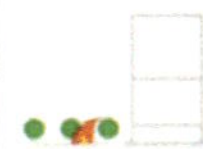
	408.3 Roof Surfaces			Not less tan 75% of the roof surfaces (zones 1 through 3, *IECC*) covered with a vegetative roof		(P) Fuel loads, fire spread (S) May difficult fire services operations	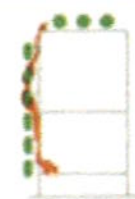
				408.3.2 Vegetative roofs Plantings selected based on their hardiness zone classifications Nonvegetated clearances in accordance with the IFC			
	409 Site Lighting						
	409.1 Light pollution control			When applicable, uplight, light trespass and glare shall be limited for exterior lighting		(P) Fire Services intervention at night could be affected	
5 Material Resource conservation and efficiency	501 General						
	502 Construction Material Management						

LEED:

Categories/ Chapter	Section/ Assessment Issues	Aims	Credits/ Scores	Requirements/ Criteria	Procedures	Fire Hazard Primary (P) & Secondary (S)	Summary
			2	Option 2:	Have 25% or 50% of parking spaces are under cover	(P) Suggestions are vegetated green roof and solar panels that may lead to increased fire risk	
	SS Credit 7.2: Heat Island Effect—Roof (1 Max Point)	To reduce heat islands1 to minimize impacts on microclimates and human and wildlife habitats.	1	Option 1:	Use of roofing materials with a solar reflectance index equal to or greater than calculated values for a minimum of 75% of roof surface		
			1	Option 2:	Install vegetated roof that covers 50% of roof	(P) Fuel loads, fire spread (S) May difficult fire services operations	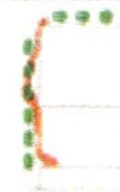
			1	Option 3:	Install high-albedo and vegetated roof surface that meet calculated values	(P) Fuel loads, fire spread (S) May difficult fire services operations	
	SS Credit 8: Light Pollution Reduction (2 Max. Point)	To minimize light trespass from the building and site, reduce sky glow to increase night sky access, improve nighttime visibility through glare reduction, and reduce development impact on nocturnal environments.	2	Option 1:	Reduce input power to interior luminaries (excluding emergency lights) with direct line of sight to any opening by 50% from 11 pm to 5 am.	(P) Poor external lighting levels may influence Fire Services intervention at night	

Structural safety/Wind design

N/A.

Irrigation

N/A.

Maintenance/Cost Estimation

N/A.

CHAPTER 12

Final considerations

About fire risk/fire safety

In terms of designing vegetative systems for fire resistance and fire safety, the overall information and requirements provided in the analyzed documents are homologous, with varying levels of detail. The majority of documents that consider fire safety precautions consider the following factors: specific design requirements of each roof layer, the introduction of firestops/firewalls in precise roof locations (i.e., gravel strips, areas free of vegetation, walls in no combustible materials), recommending regular maintenance to remove dead or excessive parts of plants, providing irrigation to maintain and keep the system alive. That being said, some differences between the documents have become apparent.

In European standards, it appears that green roofs (extensive and intensive) are considered a fire resistive system as long as they are designed in accordance with standards and regulations and are regularly maintained. These standards address concerns regarding fire safety from the design and construction of the system to when it is fully developed and functioning.

The FLL guidelines for green roofing are the most widely used in Europe and the requirements are widely considered the most reliable and successful. In the standard, all intensive green roofs are classified as hard roofs, or roofs that are adequately resistant to sparks and radiant heat. While intensive roofs are considered suitable systems that provide acceptable fire safety, extensive green roofs are only considered fire resistant if they meet specific requirements, regarding firebreaks (with precise dimensions and locations), substrate characteristics (values of minimum depth and maximum organic content), type of vegetation used, and preventing the system from drying out (required to keep the roof alive and maintained according to standard).

The success of the enforcement of these requirements is emphasized by the fact that there have been no fire accidents on green roofs that follow the FLL guidelines, since its inception almost 40 years ago. Also, it is important to highlight that many other standards, including other documents that

Green Roofs, Facades, and Vegetative Systems
ISBN 978-0-12-817694-8
https://doi.org/10.1016/B978-0-12-817694-8.00013-3

were analyzed in this research, have adapted similar requirements as FLL in terms of fire design.

Unlike FLL, UNI 11235 does not mention the fire risk topic, but it provides extensive information regarding the maintenance of green living technologies, which is an integral aspect to ensuring fire safety in these systems. The UNI 11235 standard considers the regular care of the plants (including removal of excessive or dead organic material), the roof, and the irrigation system as necessary and sufficient actions to guarantee fire resistance.

Due to the possible prolong drought in Mediterranean climates, the irrigation system is strongly recommended also for extensive green roofs. In conjunction with its maintenance, the presence of a functioning irrigation system represents a guarantee in case of fire. The SM 3700:2017 Green Roofs mentions the topic of fire risk/fire safety in several sections, providing many specific recommendations and instructions.

The documents produced in the United Kingdom (i.e., CIBSE Guideline, GRO Guideline, and Fire Performance of Green Roofs and Walls) provide similar information to that of the FLL guidelines, specifically emphasizing that green roofs are considered fire resistive when designed in compliance with the standards and are regularly maintained. The CIBSE Guidelines specify that, even though there is evidence that plants (particularly well-irrigated plants with a substrate saturated in water) can help slow the spread of fire to and from a building through the roof, dry plants can still present a fire hazard. Also, Fire Performance of Green Roofs and Walls emphasizes this point by indicating that roofs and walls that are kept moist are resistant to ignition but can present more of a fire risk if the systems dry out, which can occur during a drought period if no irrigation is provided.

To support this statement, the Fire Performance of Green Roofs and Walls research provides the results of fire tests conducted on green roofs and walls. No ignition occurred in any described cases in roofs or walls, even when completely dry, which is a very unlikely condition in a living system. The only condition that presented any fire hazard was when the majority of roof substrate was leaf mold in high concentration, which is an extremely rare condition that is below the level of maintenance required according to standards.

This is an important point because, although green living systems could not ignite or help the spread of fire, it does not completely rule them out as being a potential fire hazard, in specific (but rare) situations. This is a distinct difference between some of the documents produced in the United

Kingdom and FLL, where it specifies that all intensive roofs are adequately resistant to sparks and radiant heat.

The standards from United States provide similar requirements to the FLL for green roofs, and the research studies also present some additional interesting information and fire concerns.

The ANSI/SPRI VF—1 External Fire Design Standard for Vegetative Roofs standard provides a full list of requirements for succulent-based and grass-based green roof systems fire design including dimensions of fire stops, rules for large green roofs, reference standard for fire tests. There is no discrepancy between this standard and FLL, in relation to green roofs.

While FM Global Property Loss Prevention Data Sheets 1-35: Green Roofs Systems provides the same information regarding dimensions for border free zones and vegetation type, it also indicates that exterior fire exposure for green roofs is not possible, and interior fire exposure of green roofs shall be treated the same as conventional roofing systems on a metal deck.

Bridging the Gap: Fire Safety and Green Buildings - A Fire and Building Safety Guide to Green Construction highlights the importance of evaluating the selection of vegetation, when designing green roofs for exterior fire exposure. For example, to enhance the roof's fire performance, plants with high moisture content, such as low-growing succulents, are encouraged, while plants with high-level oils or resins should be avoided, especially in areas prone to wildfires. Moreover, the accessibility requirements are emphasized, due to the fact that, even though the green roofs may not present a direct fire threat, they cannot contribute to hindering access to fire services or the emergency operations.

Fire Safety Challenges of Green Buildings: Final Report classifies the potential fire hazards, the respective mitigation strategies, and effects if the mitigation strategies are not employed for different green living technologies. For example,

- Vegetative roofs:
  - A moderate fire threat and can contribute to fire loads, spread of fire, impact firefighter operations, impact smoke/heat ventilation, or threaten the stability of a structure during a fire event;
  - These threats can be mitigated by assuring that there is adequate fire department access, ventilation, approved materials are used, and so on;
  - If these strategies are not employed, flames can spread and firefighter access and operations can be restricted.

- Exterior vegetative covering (i.e., façades and walls):
  - A low fire threat, but can impact firefighter operations and may be susceptible to WUI (wildland–urban interface) fires;
  - These threats can be mitigated by limiting the volume of exterior vegetative covering;
  - If the mitigation strategies are not employed, this could impact the evacuation of building occupants.

This research also provides a list of 20 important fire accidents, with full references describing their characteristics and causes, but no green roofs or façades/walls were involved in any of the case studies.

Following the analysis of the documents, some final considerations and conclusions can be made:

(1) There are not reasonable grounds to consider green roof systems (extensive or intensive) as a fire risk *if designed, built, and maintained in accordance with the standards*. In several documents, green roofs and façades/walls are defined as very resistant to ignition if kept moist (which is the normal condition to keep the systems alive). It is accepted that vegetation, which receives regular maintenance and irrigation, do not pose a fire hazard. Furthermore, no noteworthy fire accidents have been caused or made worse due to green roofs.

(2) Some of the analyzed documents suggest that green roofs could be a fire hazard, but these statements are not always explained.

(3) The following table summarizes the information about fire hazards and the possible mitigation strategies for green roofs and façades/walls from all the analyzed documents:

Fire risks	Mitigation strategies
• No maintenance (the presence of dead plants, dry leaves/branches, etc.); • No irrigation (some extensive green roofs or malfunctioning irrigation systems); • High amounts of organic content in the growing media (not up to standards); • Presence of combustible elements/supports (i.e., wood).	• Obligation to create a maintenance plan; this would include a lot of effective actions against possible fire risks (removing dead plants, dry leaves/branches, etc.); • Obligation to also include an irrigation system for extensive green roofs; • Reduce the organic content in the growing media to less than 20% (according to standards); • Growing media depth is greater than 3 cm (according to standards);

Fire risks	Mitigation strategies
	• Use of fire breaks (nonvegetated borders and high wall barrier according to standards); • Use of plants with high moisture/ low resin content, instead of grasses and mosses; • Nonvegetative elements/supports must be nonflammable material; • In all cases, secure roof access provisions for firefighters.

(4) Specific documentation about fire tests on green roofs has not been extensively analyzed, but it is reported in the technical literature that many of the specified requirements are based on a large research testing program performed in 1988 by the Research and Material Testing Laboratory of Baden-Württemberg, Germany. These tests showed that extensive green roofs are nearly impossible to set on fire, and fire risk is 15–20 times greater on conventional roofs with bituminous waterproofing membranes. New tests that have been recently conducted have shown similar results. That being said, it is important to underline that there are no existing fire testing procedures for green roofs known/approved at the international scale. This gap should be addressed.

(5) Two layers of green roof systems can be considered crucial for fire safety concerns:

a. *Vegetation support course* (*substrate*): A mixture of several materials, mostly inert (i.e., volcanic stones, broken bricks, various mineral aggregates, etc.), and the remainder is organic material (i.e., topsoil, compost, etc.). It is important that the amount of organic content in the substrate is in accordance with the standards for agronomic reasons, as well as to reduce its capacity to support combustion in dry conditions. Its limit is fixed at 20%.

b. *Vegetation*: Plants can burn differently depending on the species selected (different plants can have varying fire risk depending on moisture and oil/resin content), the shape/dimension (big shrubs and trees may be more exposed to direct flame attack and could allow more fire spread compared with smaller plants), and the level of maintenance that can be provided to the plant (dry, unmaintained plants can pose fire risk and, depending on the species, some plants require vigorous maintenance activities, while others require minimal upkeep). In general, to lower any possible risk of fire on green roofs, dry foliage/branches and dead plants must be removed as part of ordinary maintenance; plants with high oil/resin content should be avoided; fast-growing plants and plants characterized by high biomass should be carefully considered in the maintenance planning phase (and possibly should be avoided).

(6) Irrigation systems, a common aspect of vegetative systems, represent the first available device for keeping the systems moist and, in some cases, for controlling possible flames. Although extensive green roofs are designed to be drought-tolerant, an irrigation system should still be installed to prevent the layering from drying out.

(7) Considering the importance of maintenance for the vegetative systems (i.e., removing dry leaves and dead plants, guaranteeing the irrigation system is properly functioning, avoiding the system from drying out, etc.), there may be an obligation to update the maintenance plan, depending on the needs of the green system and how it develops following its installation.

(8) No standards exist on green façades/walls, so standards and mandatory requirements regarding green façades/walls have to be created, including information regarding fire safety concerns of green living technology on vertical surfaces. This is a developing technology that is becoming increasingly popular, especially on tall buildings, so it is an important area that should be addressed. This is a main gap in existing standards which is highlighted by this research.

(9) No existing instructions are provided by any current standards regarding the use of vegetative systems in interior environments.

(10) Considering the doubts raised in several documents, the realistic fire risks are not fully provided in the standards; there are several questions that need to be clearly answered:

When do vegetative systems have potential flame spread capabilities?

What is the limit between fire security conditions and minimum fire risk level?

What do green living roofs and green façades have in common for concerns regarding reactions to fire?

Is it possible to provide a fire resistance rating or other classification for vegetative systems?

How do innovative vegetative systems façades/walls react to fire?

(11) From some of the analyzed documents, vegetative systems may constitute a fire hazard in specific conditions (i.e., lack of maintenance, presence of dry leaves/branches or dead plants, dry out systems, etc.). A full range of testing, conducted for every scenario (i.e., different support courses, different types of vegetation, different locations including interiors), could dispel any doubt regarding the fire risk of vegetative systems.

(12) New research and fire test campaigns are needed for specific characteristics that have not been examined to determine more detailed information on the fire performance of vegetative systems and possible new requirements that need to be introduced in the standards.

About structural safety/wind design

In general, the analyzed documents reference national standards for concerns regarding waterproofing membrane design and calculation of wind loads, but the wind design topic is approached in various ways depending on the detail and scope of the document.

The FLL guidelines frames the problems with green roofs when designing for wind loads, providing requirements for different conditions and factors that may affect the wind design, including when the waterproofing membrane is not rigidly affixed; considerations for edges and corners of the roof (which are generally more susceptible to wind); and the types of ballast to use in different case (i.e., just the substrate layering vs. increasing the weight with gravel/slabs). The FLL guidelines also references other standards for the calculation of necessary loads for wind suction, for the coefficient of wind actions, and for requirements for roofs with waterproofing membranes. Detailed values of the loads of residual moisture in soil, water storage layer, and vegetation are provided in the appendix.

The guidelines also lists several factors to consider when calculating the weight of gravel or concrete slabs if needed, including the coarseness of the vegetation; the load generated by residual moisture in the substrate; the load generated by vegetation; the bonding of the layers through the action of the roots compared with loose materials; and the wind permeability of vegetation which diffuses the pressure difference between the top and the bottom, reducing the loads.

Specific considerations are given to prevent green roofs against slipping, shearing, and erosion. Finally, the techniques for ensuring the stability of trees and shrubs are explained (bracing and anchoring the plants).

Additionally, UNI 11235 provides

- A number of factors to consider in the selection of appropriate vegetation in windy sites, including characteristics of the foliage, height of the vegetation, anchoring capacity of the root system, the stem and branches elasticity, the impact of surroundings buildings, and the potential need for temporary or permanent anchoring devices;
- A description of plant anchorage (i.e., the maximum angle of the fixings should be 60° and the calculations for wind actions should use a 1.5 multiplier as the factor of safety);
- A definition of ballast zones, including the requirements regarding the minimum width around opening and the locations of ballasts;
- References to other standards to determine the wind force and wind load estimations (Eurocode 1: Actions on structures—Part 1-4: General actions—Wind actions);
- Instructions for trees, including the vertical stability of trees in a shallow substrate must be increased in particular roots growing conditions, such as flat root ball cases; tall trees must be anchored to the building; and a list of forbidden plant species.

The SM 3700:2017 Green roofs provides several instructions for structural design, including reference to other standards for wind loads. The topic of the installation of furnishing and its safety is deepen.

ANSI/SPRI VF-1 is exclusively dedicated to wind design of extensive green roofs and it provides a complete method for the design of three roofs layering options and with two ballast options, depending on weight. Factors to determine which design and ballast option to use include the maximum wind speed, surrounding surface roughness, building height, and the parapet height. This standard provides an easy method to design extensive green roofs safe from effects of strong wind loads, but the limited choices mean that the flexibility of design is limited to the systems explained and some systems are excluded from this document (i.e., intensive green roofs and green façades/walls).

FM Global Property Loss Prevention Data Sheets 1-35: Green Roofs Systems provides several instructions and restrictions regarding the use of vegetation support course (substrate) as a ballast against wind uplift pressure and provides a safety factor and a particularly high minimum substrate depth, compared to other documents. This document also references to

Table 12.1 Summary table

§	Document name	Roof/ Façade	Fire Risk/Fire Safety	Structural safety/Wind design	Irrigation	Maintenance/Cost Estimation
STANDARDS						
4.1	UNI 11235:2015, Istruzioni per la progettazione, l'esecuzione, il controllo e la manutenzione di coperture a verde (Criteria for design, execution, testing, and maintenance of roof gardens), 2015	Roof	**N/A**	**Wind requirements are discussed in the following sections:** • §5.2 Anchoring Systems • §5.4.9 Wind Bracing Systems • §5.4.10 Anti-Erosion Layer • §5.4.13.3 Use of Ballasts • §Appendix B.5 Trees **The following summarizes the contents of this section:** • The design for temporary and permanent anchoring systems for plants under strong prevalent winds is overviewed;	**Irrigation requirements are discussed in the following sections:** • §5.4.14 Designing the Irrigation • §5.3.4.1 Descriptive Parameters of the System **The following summarizes the contents of these sections:** • §5.4.14 gives specific irrigation standards for traditional gardens and suggests irrigation types (i.e., sprinkler irrigation, drip irrigation, or subirrigation), based on the vegetation needs and the site	**Maintenance requirements are discussed in the following sections:** • §5.3.5 Level of Vegetation Maintenance • §11 Maintenance • §11.5.3 Further Instructions **The following summarizes the contents of these sections:** • §5.3.5 expresses that green roof maintenance is defined according to green roof type (i.e., extensive vs. intensive), area, frequency of maintenance, and how long it takes;

Continued

Table 12.1 Summary table

§	Document name	Roof/ Façade	Fire Risk/Fire Safety	Structural safety/Wind design	Irrigation	Maintenance/Cost Estimation
				• The evaluation of plant characteristics (i.e., foliage characteristics, vegetation height, root system anchoring capacity, and elasticity of branches and stems) is highlighted to assist in determining the design, context analysis, and wind conditions; • The use of ballasts is explained and precise values are given; • Instructions for anti-erosion layer for pitched roofs are provided; • A list of plant species that should	conditions (i.e., altitude and wind); o The need for a specific design requirements is highlighted if a subirrigation system is used, due to the characteristics of water accumulation layers; o The size needed to accommodate each system is specified; • §5.3.4.1 highlights that irrigation systems should always be used to avoid stress or damage to the	• §11 details the maintenance operations and frequency for all aspects of green roofs and splits the maintenance tasks into four main categories: o Plant maintenance; o Drainage system maintenance; o Maintenance of the gutter, drainpipe, and waterproofing membrane; o Irrigation and electrical system maintenance; • §11.5.3 provides rules to ensure the irrigation system

				be avoided is provided.	plants if a drought or fire event occurs.	runs perfectly during all seasons.
4.2	Malta Competition and Consumers Affairs Authority, Standards and Metrology Institute, SM 3700:2017 Green Roofs—Criteria for the planning, construction, control and maintenance of Green Roofs, 2017	Roof	**Fire requirements are discussed in the following sections:** • §6.2 Structural requirements • §6.4 Site analysis • §7.4.1 Components (under §7.4 Vegetation support layer) • §7.4.7 Organic matter • Table 11 • §Fire Characteristics **The following summarizes the contents of this section:** • Attention to any reflective surfaces; • Materials of drainage course shall not be flammable; • Woody plants could pose a low to	**Wind requirements are discussed in the following sections:** • §6.1 Agents and requirements • §6.2 Structural requirements • §6.4 Site analysis • §6.5 Certification of the structure • §6.10 Design loads • §8.6 Wind loads • §8.10 Furnishing **The following summarizes the contents of this section:** • Wind is mentioned as an important factor to consider in the site analysis due to stress to plants; need to consider permanent/	**Irrigation requirements are discussed in the following sections:** • §6.13 Irrigation • §8.1.3 Additional irrigation (under §8.1 Water retention) **The following summarizes the contents of this section:** • General requirements are given; • It is recommended additional irrigation is not obligatory but is recommended in regions with low rain;	**Maintenance requirements are discussed in the following sections:** • §6.6 Access and use of Green Roofs • §10 Maintenance (and subsections) **The following summarizes the contents of these sections:** • The standard imposes a maintenance plan and the operational documents: the user manual, the maintenance manual, the maintenance program; • Maintenance is classified and detailed in three

Continued

Table 12.1 Summary table

§	Document name	Roof/ Façade	Fire Risk/Fire Safety	Structural safety/Wind design	Irrigation	Maintenance/Cost Estimation
			medium risk of fire threat; • Details regarding nonvegetated margins, strips, fire breaks, and sprinklers.	temporary anchoring systems; • Certification of the structure by a qualified engineer is mandatory; • Wind loads in general terms are mentioned and reference to other standards are given; • Instructions for furnishing are given.	• Tips for different irrigations types are provided.	types: initial maintenance, long-term maintenance, and unscheduled maintenance; • Specific topics are deepened: inspections on irrigation system, repairs, and controls.
4.3	ASTM E2777−14, Standard Guide for Vegetative (Green) Roof Systems, 2014	Roof	**Fire requirements are discussed in the following section:** • §5.3.9 **The following summarizes the contents of this section:** • It recommends that succulent plants are used in areas where brush fires are a threat;	**Wind requirements are discussed in the following sections:** • §5.3.8 Wind Resistance • §7.4 Wind Scour Resistance • §7.5 Soil Reinforcement **The following summarizes the content of these sections:**	**N/A**	**Maintenance requirements are discussed in the following sections:** • §5.3.1 Maintenance • §5.3.3 Longevity • §6.1.3 Maintenance Program **The following summarizes the content of these sections:**

		• Details regarding nonvegetated margins and fire breaks.	• Provides general information regarding wind damage on perimeters, corners, and leading edge systems (i.e., gravel, stone margins, etc.); • Specifies that the dimensional requirements of leading edge systems are in the jurisdiction of the appropriate governing entities of the respective area; • References ASCE/SEI 7 when determining requirements for ballasts; • Recommendations and tips for soil media and growing media by wind scour are provided.		• Suggests how a maintenance plan should be developed; • Recommends points that should be specified for maintenance in a contract document.

Continued

Table 12.1 Summary table

§	Document name	Roof/ Façade	Fire Risk/Fire Safety	Structural safety/Wind design	Irrigation	Maintenance/Cost Estimation
4.4	ANSI/SPRI VF-1, External Fire Design Standard for Vegetative Roofs, 2017	Roof	**This entire standard is dedicated to the fire design and the following summarizes the contents of this standard:** • Complete fire system requirements are provided; • Fire stops are defined and described; • Specific rules for large roof surfaces are provided; • It explains that the access to hydrants is preferable; • Specifications for the fire design of all organic content are provided;	N/A	**The following summarizes the content of these sections:** • There is not a dedicated section for irrigation, but in §5.0 Maintenance, it is recommended that irrigation should be provided; • It is stated that the need for irrigation will vary greatly because of climate and types of plants chosen.	**Maintenance requirements are discussed in the following sections:** • §5.0 Maintenance • §C5.0 Maintenance **The following summarizes the contents of these sections:** • The importance to remove excess of biomass by plants and dead materials/foliage at least once a month and checking the moisture level once a week is emphasized; • These sections are brief and stress that it is the building owner's

			• The need for special consultation in fire and wind design needed for building greater than 45.7m in height is emphasized; • ASTM E108 and UL790 are referenced for the fire testing of roof systems; • This standard only addresses succulent-based and grass-based flat vegetative green roof systems and the fire designs for other systems are not discussed.			responsibility to maintain the vegetative roofing system.
4.5	ANSI/SPRI RP-14, Wind Design Standard for Vegetative Roofing System, 2016	Roof	**N/A**	**This entire standard is dedicated to the wind design and the following summarizes the**	**N/A**	• N/A

Continued

Table 12.1 Summary table

§	Document name	Roof/ Façade	Fire Risk/Fire Safety	Structural safety/Wind design	Irrigation	Maintenance/Cost Estimation
				contents of this standard: • It provides a method for defining the wind requirements of green roof systems—utilizing adhered roofing membranes—and the type of ballast needed, based on a set of measurable conditions. Three different design options are given: roof membrane attachment, ballasted design systems for vegetative roofing systems, and protected vegetative roofing systems. Two different types of ballast are provided.		

				• The standard uses several tables and graphs to determine the appropriate green roof system, the appropriate design options, and ballast requirements, based on the following data: ◦ Maximum wind speed on the location; ◦ Building height; ◦ Parapet height; ◦ Roughness of the surrounding ground/urban area; ◦ Classification of buildings according to wind; ◦ Snow and earthquake loads.		

Continued

Table 12.1 Summary table

§	Document name	Roof/ Façade	Fire Risk/Fire Safety	Structural safety/Wind design	Irrigation	Maintenance/Cost Estimation
4.6	CSA Group, A123.24-15 Standard test method for wind resistance of modular vegetated roof assembly, 2015	Roof	**N/A**	**This entire standard is dedicated to the wind design and the following summarizes the contents of this standard:** • Details regarding the wind flow resistant test and wind uplift resistance test of vegetative roof assemblies are provided for systems where the roof and modular vegetative system are assembled together; • §4.1 **(p. 8)** explains the significance of these tests and emphasizes their usefulness in determining if a green roof system is suitable for a particular climate;	**N/A**	**N/A**

				• Suggestions for the procedure of the tests and when to terminate the tests are made, as well as recommendations for apparatus, specimen, and instrumentation are needed.		
4.7	ASTM E2400/ E2400M−06(2015)e1, Standard Guide for Selection, Installation, and Maintenance of Plants for Green Roof Systems, 2015	Roof	**N/A**	**N/A**	**Irrigation requirements are discussed in the following section:** • §8.1 Irrigation **The following summarizes the content of this section:** • Amount of watering needed during initial phases of vegetative development and after the establishment period; • Suggests that an irrigation system is always in place, in	**Maintenance requirements are discussed in the following section:** • §8.2 Plant Maintenance **The following summarizes the content of this section:** • Suggests the amount of herbicides, insecticides, fungicides, and fertilizers needed for different vegetative types and the effects it will have on the roof membrane;

Continued

Table 12.1 Summary table

§	Document name	Roof/ Façade	Fire Risk/Fire Safety	Structural safety/Wind design	Irrigation	Maintenance/Cost Estimation
					case a drought period occurs; • Explains the difference between passive and active irrigation systems.	• Recommends the removal of undesired plants, weeds, and woody species, such as poplar, so the desired vegetation obtains full coverage and roof materials are not affected by roots; • Gives suggestions to remove weeds for projects with different levels of funding available for maintenance.
DATA SHEETS FROM INSURANCE COMPANY						
5.1	FM Global Property Loss Prevention Data Sheets 1–35: Green Roof Systems, 27 pages. April 2011	Roof	**Fire requirements are discussed in the following section:** • §2.2.12 Fire Exposure **The following summarizes the contents of this section:**	**Wind requirements are discussed in the following sections:** • §2.2 Construction and Location o §2.2.1 Building Height Restrictions	**Irrigation requirements are discussed in the following section:** • §2.3.1 Irrigation (subsection of §2.3 Operation and Maintenance) **The following summarizes the**	**Maintenance requirements are discussed in the following sections:** • §2.3 Operation and Maintenance o §2.3.2 Fertilization, Pesticides, and Plant Care

		• Suggests the evaluation of green roof systems for interior fire exposure (in the same manner as a conventional roofing system, as exterior fire exposure cannot be evaluated, due to numerous variables) (i.e., plant type, plant condition, roofing materials, etc.); • Requirements for "border zones" are given for different types of areas free of vegetation on the roof (i.e., skylights, machine rooms, antennas, etc.).	○ §2.2.2 Wind Speed Restriction ○ §2.2.3.2 Wind Uplift, Roof Ballast, and Safety Factor **The following summarizes the content of these sections:** • §2.2.1 suggests acceptable materials to use in border zones for buildings over 150ft (46m); • §2.2.2 suggests to only install the roof systems discussed in this standard in areas where the basic wind speed is less than 100mph (45 m/s); • Using corresponding data sheets (Data Sheet 1−28 and 1−29),	**content of this section:** • Irrigation solutions, such as rooftop hose bibs, allow for irrigation only during initial plant development and during drought periods, which is a suitable solution for green roof systems that do not require regularly scheduled irrigation (i.e., most extensive systems); • Intensive and simple intensive systems require a permanent irrigation solution, as they require regular irrigation.	○ §2.3.3 Leak Test and Inspection ○ §2.3.4 Warranty **The following summarizes the content of these sections:** • §2.3.2 recommends the amount of fertilization needed and suggests consulting with a roof manufacturer before determining which fertilizers or pesticides to use, as it could damage or void the warranty of the waterproofing; • Biannual inspections and selective weeding are suggested as part of the general maintenance work;

Continued

Table 12.1 Summary table

§	Document name	Roof/ Façade	Fire Risk/Fire Safety	Structural safety/Wind design	Irrigation	Maintenance/Cost Estimation
				suggestions for the roof membrane system, ballasts, and precultivated vegetated mats are made in §2.2.3.2; • §2.2.3.2 further suggests the avoidance of woody vegetation when wind uplift pressure is greater than or equal to the uplift pressure when there is a basic wind speed of 110mph (49 m/s) at a 15ft (4.5m) elevation (or less than 15ft when there is a ground roughness B).		• §2.3.3 provides instructions for the water leak test; • §2.3.4 describes the warranties involved in the vegetation and waterproofing.
GUIDELINES						
6.1	FLL (Forschungsgesellschaft Landschaftsentwicklung Landschaftsbau e.V.—The Landscaping and Landscape	Roof	**Fire requirements are discussed in the following sections:** • §8.9 Fire prevention measures	**Wind requirements are discussed in the following sections:** The structural safety and wind topic are developed crosswise	**Irrigation requirements are discussed in the following section:** • §7.9 Irrigation	**Maintenance requirements are discussed in the following section:** • §15 Maintenance services for the

	Development Research Society e.V.), Guidelines for the Planning, Construction and Maintenance of Green Roofing, 2018		• §8.6.2.2 Marginal strips
			The following summarizes the contents of this section:
			• These guidelines dictate that green roofs are "hard roofing" (adequately resistant to sparks and radiated heat) if:
			  ○ Intensive green roofs are considered as roofs that are resistant to flying sparks and radiant heat;
			  ○ Extensive green roofs are resistant to flying sparks and radiant heat if they have a list of characteristics;
			  ○ In addition, during construction and use, general rules
				to all the guidelines, referring to different layers/components and different aspects of designing green roofs:
				• §5.3 Determination of site conditions for vegetation
				• §7 Requirements for construction and building materials
				• §7.6 Load assumptions
				• §7.7 Protection against falling
				• §8 Structural requirements (and subsections)
				• §9.3 Water retention
				• §14.3 Securing the stability of woody plants
				• Appendix A
				The following summarizes the content of these sections:
					• §9.4 Water storage and additional irrigation
					The following summarizes the content of this section:
					• Irrigation shall be provided by at least one water connection on the roof;
					• It asks to reference the provisions laid out in DIN 1717, to determine the number of water connections, number of pipes, and water pressure requirements.
						development and maintenance of vegetation**The following summarizes the content of this section:**
						• It specifies that the development care and maintenance for ground-level greening in DIN 18919 can also be applied to intensive greening;
						• The care for simple intensive greening and extensive greening has to be dictated by the greening method, vegetation type, and development of the vegetation;
						• For all green roofs, it is recommended that contractual arrangements are made for an extended period

Continued

Table 12.1 Summary table

§	Document name	Roof/ Façade	Fire Risk/Fire Safety	Structural safety/Wind design	Irrigation	Maintenance/Cost Estimation
			of fire prevention are provided for construction phase and use. • Marginal strips in gravel are introduced as elements useful for fire prevention. • As in all the standards, the maximum organic matter of the substrate is 20%.	• §7.6 Load assumptions provides basic important information to consider in design phase (i.e., maximum water capacity, loads of plants, other structural elements, point loads); • Measures for protections against falling; • The marginal strips in gravel are introduced as required elements against wind suction, besides inspection areas to waterproofing membrane, and transitions to rising structural element (marginal strips may be considered functional to fire prevention);		with a professional (i.e., landscape architect or contractor), with maintenance conditions clearly defined, such as the objective of the greening (i.e., aesthetically orientated garden vs. ecologically optimized); A list of maintenance tasks, the frequency that these tasks must be performed, and tips for serving the vegetation are provided.

				• Reference standards are provided for negative wind pressure; • §8.10 Protection against material displacement deepens problems related to surface erosion, slipping, and exceeding the angle of repose; • Installation of furnishing is explained in detail; • Requirements for the bracing and anchoring of trees and large shrubs, including, but not limited to, requirements for the volume of substrate, maintenance during the bracing/ anchoring period, materials, and framing shapes.		

Continued

Table 12.1 Summary table

§	Document name	Roof/ Façade	Fire Risk/Fire Safety	Structural safety/Wind design	Irrigation	Maintenance/Cost Estimation
6.2	CIBSE Knowledge Series 11, KS11 Green Roofs, 2007	Roof	**Fire requirements are discussed in the following section:** • §3 Design Considerations for Green Roofs **The following summarizes the contents of this section:** • Recommends pebble barriers are used for fire breaks and to follow the FLL standard; • Suggestions for the substrate depth, regularity of fire breaks, and vegetation barriers are made for extensive roofs to be considered unlikely to be ignited by sparks.	**N/A**	**Irrigation requirements are discussed in the following section:** • §3 Design Considerations for Green Roofs **The following summarizes the contents of this section:** • Recommendation for watering requirements is made for intensive, semi-intensive, and extensive green roofs, as well as green roof systems that are south facing or in dry UK climates; • Suggests that watering is required for all green roofs (excluding bio-diverse) when there is a drought of 6 in the United Kingdom.	**Maintenance requirements are discussed in the following sections:** • §5 Maintaining Green Roofs • §6 Retrofitting Green Roofs • §7 Costs and Financial Considerations **The following summarizes the content of these sections:** • Several short suggestions are made regarding maintenance, retrofitting, and cost.

6.3	CIBSE Guidelines for the Design and Application of Green Roof System, 2013	Roof	**Fire requirements are discussed in the following section:** • §6.3 Fire Precautions **The following summarizes the contents of this section:** • Recommends pebble barriers are used for fire breaks and to follow the FLL standard; • It emphasizes that green roofs can prevent spread of fire threats, but in contrast, can pose more of a threat if the plants are dry; • Recommendations for openings in the green roof, regularity of fire breaks, and irrigation of fire breaks are made; • The level of fire risk for extensive vs. intensive green roof systems is explained.	**Wind requirements are discussed in the following sections:** • §5.3 Wind Design o §5.3.1 Wind Uplift and Ballast o §5.3.2 Calculation of Wind Uplift Design Pressures **The following summarizes the content of these sections:** • The fact the corners and edges have a much higher wind pressure than the central area is emphasized; • It also emphasizes that tall structures have a greater risk of wind uplift and ballasting requirements to vary depending on height and wind speed; • SPRI (2010a) is referenced for designing wind	**Irrigation requirements are discussed in the following section:** • §7.5 Irrigation **The following summarizes the contents of this section:** • Very general irrigation suggestions are made in this section.	**Maintenance requirements are discussed in the following sections:** • §7 Maintenance Issues • §8.3 Costs of a Green Roof System **The following summarizes the content of these sections:** • §7 provides details regarding factors that will affect the level of maintenance required, such as height, microclimate, soil type, soil depth, irrigation, species, access, and stage of maintenance (i.e., during installation/construction, during development of green roof systems, ongoing maintenance requirements);

Continued

Table 12.1 Summary table

§	Document name	Roof/ Façade	Fire Risk/Fire Safety	Structural safety/Wind design	Irrigation	Maintenance/Cost Estimation
				uplift resistance for green roofs; • Suggestions for the safety factor of the wind uplift calculations are made, depending on different conditions.		• Further details are provided for the procedures for fertilizing, irrigation, weeding, pruning, replanting, and maintenance of waterproofing membrane; • Suggestion for the regularity of maintenance is made for each green roof type (i.e., extensive, intensive, etc.); • §8.3 provides a brief overview of costs related to green roofs and their maintenance.
6.4	Groundwork Sheffield, The GRO Green Roofs Code, 2014	Roof	**Fire requirements are discussed in the following section:** • §3.5 Fire **The following summarizes the contents of this section:**	**Wind requirements are discussed in the following section:** • §3.2 Structural design • §3.2.1 Wind **The following summarizes the**	**Irrigation requirements are discussed in the following section:** • §3.6 Irrigation **The following summarizes the**	**Maintenance requirements are discussed in the following sections:** • §5 Green Roof Maintenance

		• Suggestions to increase fire resistance are made; • Requirements for fire breaks and nonvegetated strips are explained, in order to reduce fire risk on extensive green roofs; • Emphasizes that intensive green roofs can be considered "hard roofs" if they have a deep substrate and are regularly maintained and irrigated.	**content of this section:** • General instructions are provided for designing a green roof as a ballast to prevent items not mechanically fixed from wind uplift; • General instructions are provided for ballasts and erosion control measures are made for areas with high wind suction loads (i.e., edges and corners); • List of the standards to refer for structural design.	**contents of this section:** • Other standards are referenced for irrigation requirements; • Irrigation during the initial development of the green roof is recommended, but irrigation can be reduced or avoided, depending of the green roof system used.	• §5.1 General Maintenance Actions • §5.2 Maintenance Actions by Roof Type **The following summarizes the content of these sections:** • Other standards are referenced for general landscaping and maintenance; • Explanations of four types of maintenance actions are provided: irrigation, fertilization, plant management, and general clearance/removal; • The regularity for the maintenance actions to be performed is described for different types of roof systems (i.e., extensive, semi-intensive, intensive, and bio-diverse).

Continued

Table 12.1 Summary table

§	Document name	Roof/ Façade	Fire Risk/Fire Safety	Structural safety/Wind design	Irrigation	Maintenance/Cost Estimation
6.5	State of Victoria through the Department of Environment and Primary Industry, Growing Green Guide: A guide to green roofs, walls and Façades in Melbourne and Victoria, Australia, 2014	Roof	**Fire requirements are not discussed, but are briefly mentioned in the following sections:** • §2.6 Nearby Vegetation • §3.6 Planning, Regulations, and Local Laws • §5.6 Growing Substrate • §8.2 Maintenance Tasks **The following summarizes the contents of these sections:** • The removal of dead or dry vegetation is suggested as part of the regular maintenance of green roofs, to prevent fire hazards; • The installation of breaks in vegetation	**Wind requirements are discussed in the following sections:** • §2.1 Climatic Factor On-Site • §5.11 Wind Consideration • §5.12 Slopes and Wind Protection **The following summarizes the content of these sections:** • Encourages an understanding of the local climate, wind load, and the influence wind at high levels can have on temperature and the dehydration of vegetation before determining which plant species are suitable for the wind conditions;	**Irrigation requirements are discussed in the following sections:** • §Frequently Asked Questions: "Is Irrigation Necessary for a Green Roof, Wall, or Façade?" • §2.1 Climate Factors On-Site • §3.2 Planning for Drainage and Irrigation • §3.4 Sourcing Skill, Expertise and Information: Irrigation Consultant/ Hydraulic Engineer • §5.10 Irrigation **The following summarizes the contents of these sections:** • Recommends irrigation for green roofs, walls, and	**Maintenance requirements are discussed in the following sections:** • §3.1 Designing for Maintenance • §3.5 Cost Consideration • §3.6 Planning, Regulation, and Local Laws • §8 Maintenance o §8.1 Maintenance Planning o §8.2 Maintenance Tasks **The following summarizes the content of these sections:** • §3.5 explains the different costs involved in the maintenance of green roof/walls/ façades, including

		and plants with low biomass is suggested for fire resistance.	• Suggestions are made for the waterproofing membrane, ballasts, and pitched roofs to combat wind uplift.		
	Façade	**Fire requirements are not discussed, but are briefly mentioned in the following sections:** • §2.6 Nearby Vegetation • §3.6 Planning, Regulations, and Local Laws • §5.6 Growing Substrate • §8.2 Maintenance Tasks **The following summarizes the contents of these sections:** • The removal of dead or dry vegetation must be	**Wind requirements are discussed in the following section:** • §Definitions **The following summarizes the content of this section:** Suggests that twining climbers are used over plants and that foliage type and size is carefully considered, in order to avoid significant problems for green façades, particularly at height.	façades, even though some systems may be considered "drought resistant"; • Explanation of different irrigation systems, along with their delivery method, advantages, and disadvantages, is provided.	water for irrigation, fertilizer, replacement plants, weeding, pest and disease management, periodic inspections, retensioning of cables, repair of loose fixings in the irrigation system, and access costs (if an elevated work platform is required); • The other sections make suggestions for actions to be considered in the planning of maintenance, including the objectives based on the design intent, the performance targets (i.e., time frame for complete coverage of

Continued

Table 12.1 Summary table

§	Document name	Roof/ Façade	Fire Risk/Fire Safety	Structural safety/Wind design	Irrigation	Maintenance/Cost Estimation
			part of the regular maintenance of green walls or façades, to prevent fire hazards; Vegetation on green walls that are regularly irrigated and maintained and green façades with deciduous or evergreen climbers that have dry leaf litter regularly removed are unlikely to pose a fire risk.			vegetation), responsibilities of personnel, the training requirements and safety measures, and necessary resources; • The development of a contract is suggested, with a schedule for regularity of maintenance, the specific maintenance task involved, and the role of the maintenance contractor.
MANUALS						
7.1	Green Wall 101: Systems Overview and Design, Green Roofs for Healthy Cities, 2014	Façade	**N/A**	**Wind requirements are discussed in the following section:** • §5.1.1 Site Analysis **The following summarizes the**	**Irrigation requirements are discussed for each class of green façade, including**	**Maintenance requirements are discussed in the following sections:** • Green Façades (Modular Trellis

				content of this section: • Wind force is considered an "environmental factor" that affects green walls, but no specific requirements are given.	**the following sections:** • Green Façades (Modular Trellis Panel System, Wire-Rope and Cable Net Systems) §6.1.3 Water Supply and Drip Irrigation • Living Walls §6.2.2 Water Supply and Drip Irrigation **The following summarizes the contents of these sections:** • §6.1.3 emphasizes the need for an irrigation system, especially in south- and west-facing walls, as they are exposed to more sun; • It is explained that the irrigation system selected (i.e., automatic irrigation, low-flow bubblers, and low-angle strip	Panel System, Wire-Rope and Cable Net Systems) §6.1.7 Maintenance Requirements • Living Walls §6.2.7 Maintenance Considerations • §5.3 Budget Development ○ §5.3.1 Cost Estimation **The following summarizes the content of these sections:** • §6.1.7 explains some of the factors that need to be taken into account when developing a maintenance plan for green façades, including requirements for irrigation, soil, access to vegetation, maintenance workers (training and cost), the system to

Continued

Table 12.1 Summary table

§	Document name	Roof/ Façade	Fire Risk/Fire Safety	Structural safety/Wind design	Irrigation	Maintenance/Cost Estimation
					spray systems) depends on a number of factors, including type of plant, water availability, sun exposure, design goal, etc.; • In §6.2.2, requirements for the amount of irrigation, regularity of irrigation, and required systems are explained for indoor and outdoor living walls; • Recommendations for systems are made to prevent the failure of the living wall systems and provide adequate irrigation (i.e., backflow preventers, water quality monitoring systems, drip irrigation systems,	accommodate growing vegetative weight, and inspection of plants and underlying structure; • §6.2.7 empathizes that living walls require additional maintenance and details the additional factors that need to be considered, including pruning, replacement of dead plants, and monitoring of the plants, irrigation, plant nutrients, soil quality, diseases/pests, and the underlying system; • §5.3 and §5.3.1 examines some of the factors that will affect the cost of the green wall system, including

					synthetic moisture fabrics, etc.); • It emphasizes that the living walls are most at risk of failure if there is an interruption or irregularity in the water supply.	the size of the project, type of system used, location, availability of materials, timeline, amount of maintenance required, and labor (i.e., designers, contractors, maintenance workers).
7.2	Advanced Green Roof Maintenance, Green Roofs for Healthy Cities, 2011	Roof	**Fire requirements are discussed in the following section:** • Appendix D **The following summarizes the content of this section:** • References ANSI/ SPRI VF-1 External Fire Design Standard for Vegetative Roofs to appropriately address the topic of fire design.	**Wind requirements are discussed in the following section:** • Appendix D **The following summarizes the content of this section:** • References ANSI/ SPRI RP-14 Wind Design Standard for Vegetative Roofing Systems to appropriately address the topic of fire design.	**Irrigation requirements are discussed in the following sections:** • §7 Irrigation System ○ §7.4 Standard Seasonal Maintenance **The following summarizes the contents of these sections:** • Different types of irrigation systems are introduced (i.e., drip irrigation, overhead spray systems, and hand	**This entire document is dedicated to the maintenance and the following summarizes the contents of this handbook:** • An extensive description of the maintenance plan is provided, including every aspect that should be factored into it (i.e., nonvegetative component maintenance,

Continued

Table 12.1 Summary table

§	Document name	Roof/ Façade	Fire Risk/Fire Safety	Structural safety/Wind design	Irrigation	Maintenance/Cost Estimation
					watering) and technical aspects of each are explained (i.e., functioning, scope, use, etc.); • Different types of maintenance to be performed on the different irrigation systems throughout the year are explained (preparing for winter, start-up in spring, and during usage period).	seasonal recommendations, fertilizers/pesticides, erosion control, etc.); • An extensive description of factors that could influence the cost is also provided (i.e., size, system complexity, type of plants, ease of access, cost of labor, etc.); • The procedures for maintenance are explained, separating them into three categories: during installation, during initial development and growth of vegetation, and after the vegetation is established; • An entire section overviews how to diagnose problems

						with the roof systems and tools to use in this process; • Five case studies from Canada and the United States are provided.
BUILDING CODES						
8.1	International Code Council, IBC—International Building Code, 2018	Roof	**Fire design is mentioned in numerous chapters/sections, but the following sections pertain specifically to green roofs:** • §1507.16 Vegetative Roofs, Roof Gardens and Landscaped Roofs • [BF 1507.16.1] Structural Fire Resistance • §1607.13.3 Occupiable Roofs ○ §1607.13.3.1 Vegetative Roofs, Roof Gardens and Landscaped Roofs **The following summarizes the contents of these sections:** • §307 lists hazardous materials and vegetation as not included as a hazardous material; • §1607.13.3 and §1607.13.3.1 emphasize that the load of green roofs shall comply with specific requirements.			
		Façade	**This building code includes *Chapter 12: Interior Environments* and *Chapter 14: Exterior Walls*, but there are no explicit mentions of green façades.**			

Continued

Table 12.1 Summary table

§	Document name	Roof/ Façade	Fire Risk/Fire Safety	Structural safety/Wind design	Irrigation	Maintenance/Cost Estimation
8.2	International Code Council, IGCC—International Green Construction Code, 2015	Roof	**Fire requirements are discussed in the following section:** • §408.3.2.4 Vegetative Roofs **The following summarizes the content of this section:** • References the IFC (International Fire Code) for determining required nonvegetated clearance sizes and locations.	**N/A**	**N/A**	**Maintenance requirements are discussed in the following sections:** • Table 902.1 Commissioning and Inspection Plan Requirements • Table B101.1 Post-Occupancy Commissioning Reporting **The following summarizes the content of these sections:** • This code specifies that filed inspections and reports are required during the installation of the protective membranes, base materials, soils, vegetation, and post-occupancy according to maintenance plan.

8.3	International Code Council, IEBC—International Existing Building Code, 2012	Roof Façade	**This building code includes several sections and guidelines on fire ratings of archaic materials and assemblies, but there are no explicit mentions of green roofs or façades.**	**This building code includes *Appendix C: Guidelines for Wind Retrofit of Existing Buildings*, but there are no explicit mentions of green façades.**	**N/A** **N/A**	**N/A** **N/A**
SAFETY CODES						
9.1	NFPA 101: Life Safety Code, 2015	Roof Façade	**This fire safety guide includes the description of fire escapes, safe areas, egresses, etc., but there are no explicit mentions of green roofs or façades.**	**N/A** **N/A**	**N/A** **N/A**	**N/A** **N/A**
9.2	2015 NFPA: 5000 Building Construction and Safety Code, 2015	Roof	**This fire design guide includes §*Chapter 38: Roof Assemblies and Roof Structures*, but there are no explicit mentions of green roofs.**	**N/A**	**N/A**	**N/A**

Continued

Table 12.1 Summary table

§	Document name	Roof/ Façade	Fire Risk/Fire Safety	Structural safety/Wind design	Irrigation	Maintenance/Cost Estimation
		Façade	**Fire requirements are discussed in the following sections:** • §7.2.3.2.12 Exterior Nonbearing Walls • §7.2.5.6.7 Exterior Walls • §7.2.5.6.7 Exterior Nonbearing Walls **The following summarizes the content of these sections:** • §7.2.3.2.12 lists materials that can be used in nonbearing exterior walls and have adequate fire resistance; • §7.2.5.6.7 lists materials that can be used in exterior	**N/A**	**N/A**	**N/A**

			walls with a fire resistance requirements of 2 hours or less, including noncombustible materials, limited combustible materials, certain types of fire retardant-treated wood, and cross-laminated timber that meets specific requirements; • §7.2.5.6.7 specifies that exterior nonbearing walls are permitted if they meet the testing requirements of NFPA 285, Standard Fire Test Method for Evaluation of Fire			

Continued

Table 12.1 Summary table

§	Document name	Roof/ Façade	Fire Risk/Fire Safety	Structural safety/Wind design	Irrigation	Maintenance/Cost Estimation
			Propagation Characteristics of Exterior Non-Load-Bearing Wall Assemblies Containing Combustible Components.			
CITY CODES						
10.1	Toronto Municipal Code, Chapter 492, Green roofs, 2013	Roof	**Fire requirements are discussed in the following sections:** • §492-6 Application for Approval • §492-9.F Toronto Green Roof Construction Standard: Mandatory Provisions **The following summarizes the**	**Wind requirements are discussed in the following section:** • §492-9.E Toronto Green Roof Construction Standard: Mandatory Provisions **The following summarizes the content of this section:**	**Irrigation requirements are discussed in the following section:** • §492-9.M Toronto Green Roof Construction Standard: Mandatory Provisions **The following summarizes the contents of this section:**	**Maintenance requirements are discussed in the following sections:** • §492-3 Maintenance of Green Roofs and Health of Vegetation • §492-9.N Toronto Green Roof Construction Standard: Mandatory

		contents of these sections: • It states the "Green Roof Declaration Form" may require fire safety provisions, so this must be taken into account; • Dimensional regulations for vegetation-free border zones around roof penetrations, intersecting walls, parapets, upturns, and mechanical equipment are provided.	• The sections states that a report, stamped by an engineer, should be provided, which highlights how wind uplift pressures were calculated and how they are addressed in the design.	• It is emphasized that, in order to initiate and sustain the vegetation on the green roof, adequate irrigation measures are necessary.	Provisions **The following summarizes the content of these sections:** • §492-3 states that every green roof should be maintained according to the Toronto Green Roof Construction Standard; • §492-9.N references CSA-S478-95 (The Guideline on Durability in Buildings) to assist in identifying a sufficient amount of routine maintenance actions and inspections when developing the maintenance plan;

Continued

Table 12.1 Summary table

§	Document name	Roof/ Façade	Fire Risk/Fire Safety	Structural safety/Wind design	Irrigation	Maintenance/Cost Estimation
						• It emphasizes that the maintenance plan for the green roof systems should pertain to the specific growth and type of vegetation, address any necessary replanting, and ensure that complete coverage is achieved and maintained within three growing seasons; • It states that, in order to receive a permit for a green roof, the maintenance plan must be submitted with the building application.

10.2	Toronto Green Roof Construction Standard (TGRCS), Supplementary Guidelines	R	**Fire requirements are discussed in the following section:** • §3.1 Fire Safety **The following summarizes the contents of this section:** • It lists the OBC (Ontario Building Code) requirements that relate to fire safety; • It lists a series of best practices, including o Avoiding vegetation that increases fire risk; o Keeping vegetation away from walls, mechanical equipment, etc.;	**Wind requirements are discussed in the following section:** • §2.4 Wind Uplift **The following summarizes the content of this section:** • It lists the OBC (Ontario Building Code) requirements that relate to wind uplift; • It summarizes the best practices to determine wind uplift and references the FM Global Data Sheets, FLL, and the Wind Design Standard for Vegetative Roofing Systems (a document currently in development) for more information.	**Irrigation requirements are discussed in the following section:** • §5.3 Irrigation **The following summarizes the contents of this section:** • It lists the OBC (Ontario Building Code) and City of Toronto By-Law (No.583-2009 & §492-9 M) requirements that relate to irrigation.	**Maintenance requirements are discussed in the following section:** • §6.2 Maintenance Plan **The following summarizes the content of this section:** • It list the OBC (Ontario Building Code) and City of Toronto By-Law (No.583-2009 & §492-3) requirements that relate to maintenance (which also includes references to CSA-S478-95 "Guideline on Durability in Buildings" for development of a maintenance plan);

Continued

Table 12.1 Summary table

§	Document name	Roof/ Façade	Fire Risk/Fire Safety	Structural safety/Wind design	Irrigation	Maintenance/Cost Estimation
			o Using noncombustible cladding in areas near vegetation; o Appropriate placement of fire breaks and gas lines; o Considerations for firefighting and burning vegetation on paths of egress; • References FM Global Data Sheets for more information.			• Lists a number of factors to consider when developing a maintenance plan for a green roof during the first 2–3 years and beyond.
RESEARCH REPORTS						
11.1	Department for Communities and Local Government-London, Fire Performance of Green Roofs and Walls, 2013	Roof	**This entire report is dedicated to the fire design and the following summarizes the contents of this document:**	**Wind requirements are discussed in the following section:** • §2.2 Wind Design **The following summarizes the**	**Irrigation requirements are discussed in the following sections:** • §1.1 Green Roof **The following summarizes the**	**Maintenance requirements are discussed in the following sections:** • §4.6.4 Installation and Maintenance **The following**

			• This standard begins by describing and classifying different types of green roofs/walls; • States that green roofs/walls are generally resistant to ignition, but recognizes that they could ignite as there has been minimal fire tests on green roofs (and none for green walls); • UK Guidelines, FLL, ANSI/SPRI VF-1, FM Data sheets 1–35, and GRO are referenced to determine the primary measures used for fire prevention on green roofs (which can	**content of this section:** • This section suggests referencing FM Data sheet 1–35 and European guides for this topic.	**contents of these sections:** • It references the GRO Green Roof Code Guide and Factory Mutual Data sheet 1–35 to suggest appropriate irrigation systems for extensive, semi-intensive, and intensive roof systems.	**summarizes the content of this section:** • Maintenance of green roofs is briefly mentioned in this section.

Continued

Table 12.1 Summary table

§	Document name	Roof/ Façade	Fire Risk/Fire Safety	Structural safety/Wind design	Irrigation	Maintenance/Cost Estimation
		Façade	also be valid for green façades): o Increasing the noncombustible content; o Decreasing the amount of organic content; o Preventing the system from drying out; o Adding fire breaks for extensive green roofs. • Specific instructions regarding layering, fire breaks, and access to water from the other standards are reviewed; • Regulation requirements are	**N/A**	**Irrigation requirements are discussed in the following sections:** • §1.2 Green Walls **The following summarizes the contents of these sections:** • State that climbing plant systems usually have an irrigation system, but they can survive without one, if rooted into the ground; Suggests irrigation lines for hydroponic green walls, as the plants rely exclusively on irrigation and nutrients (as opposed to substrate or soil).	**Maintenance requirements are discussed in the following sections:** • §4.6.4 Installation and Maintenance **The following summarizes the content of this section:** Maintenance of green façades is briefly mentioned in this section.

		analyzed to determine if green roofs effect the fire risk related to internal/external fire spread, access, or functional equipment; • Two types of fire tests (cone calorimetry tests and large scale tests) are described and results are provided: o Substrate generally does not ignite, cause flame spread, or reach temperatures that would ignite materials used below; o Substrates in living walls cannot be			

Continued

Table 12.1 Summary table

§	Document name	Roof/ Façade	Fire Risk/Fire Safety	Structural safety/Wind design	Irrigation	Maintenance/Cost Estimation
			ignited, even when completely dried out (ignition only occurs when substrate is 100% dried leaf mold, which is highly unlikely)			
11.2	National Association of State Fire Marshals: Tidwell, Jim, and Jack J. Murphy, Bridging the Gap: Fire Safety and Green Buildings. A Fire and Building Safety Guide to Green Construction, 2010	Roof	**Fire requirements are discussed in the following section:** • §III.B Vegetative Roof Systems **The following summarizes the content of this section:** • Safety concerns effected by a green roof system during a fire event are emphasized, including roof loads (i.e., materials	**Wind requirements are discussed in the following section:** • §III.B Vegetative Roof Systems **The following summarizes the content of this section:** • Specifies that all necessary calculations to determine wind pressures and wind loads should be in dry conditions and	**N/A**	**Maintenance requirements are discussed in the following section:** • §III.B Vegetative Roof Systems **The following summarizes the content of this section:** • Maintenance is briefly described for each type of green roof (i.e., extensive, semi-intensive, intensive), and the

		saturated with water increasing weight of roof), ease of roof access, and the lack of fire exposure ratings; • The FM (Factory Mutual) Property Loss Prevention Data Sheets—Green Roof Systems, January 2007 are referenced for exterior green roof fire exposure and specifiy that interior fire exposure should be considered similar to traditional roofs; • It emphasizes that exterior fire exposure needs to be careful considered, especially when selecting plant types (plants with high	without a vegetation load, and particular attention should be given to areas prone to high winds; • Suggests that additional roofing materials or objects that could become flying debris should be avoided, so it seems to discourage green roofs.		importance of a maintenance plan is emphasized.

Continued

Table 12.1 Summary table

§	Document name	Roof/ Façade	Fire Risk/Fire Safety	Structural safety/Wind design	Irrigation	Maintenance/Cost Estimation
			moisture content enhance fire performance, while plants with volatile oils or resins should be avoided when considering fire exposure).			
11.3	The Fire Protection Research Foundation: Meacham et al., Fire safety Challenges of Green Buildings: Final Report, 2012	Roof Façade	**This entire report is dedicated to the fire design and the following summarizes the contents of it:** • Potential fire hazards, the level of concern, and potential mitigation strategies caused by different green living elements (i.e., vegetative roofs, exterior vegetative coverings, interior vegetation, and vegetation for	**N/A** **N/A**	**N/A** **N/A**	**N/A** **N/A**

		shading) are described; • The level of increased fire risk, if the green living elements do not undergo a mitigation strategy, is emphasized; • An assessment of international energy label requirements (i.e., LEED, BREEAM, GREENMARK, and IgCC) is provided, specifically examining the effect that suggests these requirements would have on the fire safety objectives of a building that may have not been previously considered.			

other standards (i.e., Data Sheets 1-28 Wind Design; Data Sheets 1-29 Deck Securements and Above-Deck Roof Components) for detailed instructions relating wind design calculations.

Some specific mandatory instructions are given regarding the use of pavers versus gravel, the anchoring of precultivated vegetated mats, and requirements for woody vegetation.

Following the analysis of the documents, some final considerations and conclusions can be made:

(1) Different reference standards are provided for calculating wind uplift pressure. Although the rules for defining ballasts are similar throughout the documents, there are slight differences with different standards. It is important to note that, generally, reduction of the weight of green roofs is seen as a positive to reduce overall load on building structures, but elements, such as ballasts, which can significantly increase the overall weight, are seen as necessary to ensure safety against wind, particularly on edges and corners. In cases where wind loads are particularly strong, it is suggested that unique engineering analyses are conducted, especially for intensive roofs with high plants.

(2) All necessary information needed regarding the definition of anchoring systems for trees and big shrubs to avoid uprooting is not complete in the documents that have been analyzed. Additional research on this is necessary to determine what anchors currently exist, which method of anchoring should be used, when are they necessary, and where should they be installed for woody plants.

(3) Methods to avoid debris are given, but in cases with more complex landscape designs, such as intensive greening systems with trees, no explicit instructions are provided.

(4) Wind design of green living façades is not addressed in international standards.

About irrigation

General information regarding irrigation of green roofs is explained in almost all of the analyzed documents, with specific tips for green roofs, and reference standards for any technical requirements. In most of the documents, irrigation is strongly recommended, even when draught-tolerant greening is used to avoid stress to plants in the first stages of the planting or during long dry periods. Like the selection of vegetation, the selection of the irrigation system is part of the overall landscape of a green roof and has to be designed by experts in the field, taking into account the specific needs of the plants and using the water source efficiently.

Additionally, as previously mentioned, the irrigation system can prevent green systems from drying out and is a good measure for fire prevention, so their installation may become mandatory.

About maintenance/cost estimation

The maintenance of green roofs is a key topic for green living technologies and the frequency of maintenance is the main parameter to classify if a system is an extensive, semiintensive, or intensive green roof. Extensive green roofs require a minimum maintenance, while semiintensive and intensive green roofs require more. The frequency and the type of maintenance operations performed is considered one of the main aspects to guarantee successful green living systems, in terms of longevity, fire safety, and occupant safety.

Both FLL and UNI 11235 provide the most extensive information regarding maintenance operations.

The FLL standard provides a full list of tasks and frequency that they should be preformed for extensive and intensive greening. It also refers to the DIN 18919 standard, for maintenance actions for landscaping at ground level.

The UNI 11235 is by far the most extensive and reliable standard for maintenance actions. In addition to providing the information on the operations and frequency, it separates maintenance actions into different categories (i.e., starting maintenance to control, starting maintenance for ordinary maintenance, ordinary maintenance, extraordinary maintenance, maintenance of the drainage system, gutters, drainpipes and waterproofing membrane, and maintenance of irrigation and electrical systems).

The standard also supplies information for the amount of man hours required per area (m^2) per year for different type of green systems. This is imperative to understand the amount of costs a building owner or manager will need to pay to maintain a successful green living system and necessary to create an accurate maintenance plan. An example of a maintenance plan is provided and references to another standard (UNI 11540:2014 – Guidelines for writing and realizing of a maintenance plan of roofs with waterproofing membranes) are provided, with a view to ensure a thorough maintenance plan considered from the design phase of any new green system.

The importance of the creation of a detailed maintenance plan in emphasized in the following analyzed documents:

- ASTM E2777-14 Standard Guide for Vegetative (Green) Roofs Systems, 2014.
- State of Victoria through the Department of Environment and Primary Industry, Growing Green Guide. A guide to green roofs, walls and facades in Melbourne and Victoria, Australia, 2014.
- Toronto Municipal Code, Chapter 492 Green Roofs, 2013.
- GRHC Advanced Green Roof Maintenance, 2011.

The GRHC – Green Roofs for Healthy Cities, Advanced Green Roof Maintenance, 2011 is an extensive document, dedicated entirely to maintenance of green roofs. It identifies factors that may influence costs, the specific operations during the installation of the green roof, necessary operations during the initial development of vegetation, and operations after the establishment of vegetation. Then, maintenance contracts and warranties are outlined.

ANSI/SPRI RP-14 2010 suggests that additional maintenance actions should be taken on a green roof following every strong wind event.

It is possible that there are no gaps in existing standards for maintenance concerns. Following requirements and suggestions in existing standards and documents, it appears that a thorough maintenance plan can provide a successful and safe green living system. That being said, due to the popularity of green roofs, additional considerations could be made to help ensure safe access and conditions for gardeners and other workers performing the maintenance actions: access pathways to the roof should be suitably considered by the building design team. Like the previous topics (fire safety, wind design, and irrigation), maintenance requirements for green living systems are not featured in the international codes.

Conclusions

This research highlights the requirements and practices regarding green vegetative systems, as expressed in existing standards, guidelines, research, etc.

The requirements from all the analyzed documents provide extensive information of green roof design in regard to the four main topics: fire risk/fire safety; wind design and structural safety, irrigation, maintenance/cost estimation. That being said, there are clear areas where the existing standards can be developed, including undefined fire and wind design considerations for certain green roof scenarios.

For green roofs, new requirements and recommendations may be needed to increase the understanding of green systems. Several standards use appendix and commentary sections to provide more information, but these can be further expanded, considering green systems require design strategies atypical of traditional building components, as they are living systems that are characterized by factors that are not easily definable, measurable, or classifiable. Additional recommendations and deepening specific topics can help make designers aware of possible solutions or scenarios, which can solve potential problems that may be difficult to define as precise "requirements" or "rules."

Regarding fire risk, a new test campaign, supported by new research, should be developed to clarify the doubts about the fire reaction of green living technologies raised by some of the analyzed documents. New research should investigate the following:

- The fire reaction of a full range of green living technologies (i.e., different type of vegetation, type of substrates, irrigation, presence of ventilation layer in case of facades, etc.);
- The fire reaction under a full range of scenarios (i.e., climate, accessibility, tall buildings, etc.);
- How to set realistic case scenarios to be tested;
- How to express the fire test results.

The fire tests should clarify the following:

(1) The cases/scenarios in which green vegetative systems may have a potential flame capability and the worst cases scenarios for roofs, façades, and interiors;

(2) The limiting conditions to ensure fire safety of green vegetative systems.

The output of the new test campaign, supported by the new research, should provide complete guidelines/standards to ensure the fire protection of green vegetative systems.

Also, in case of wind design, significant gaps in the standards can be noticed. A new research to investigate both green roofs and façades/walls is needed to answer the following:

(1) What plants characteristics should be avoided in design phase to ensure wind safety?

(2) Which cases/scenarios will require the use of anchoring systems?

(3) What type of anchoring systems should be chosen in different cases?

(4) What requirements and solutions are needed to avoid debris, especially in case of complex landscape design?

Similar to the required test campaign for fire safety, the output of this new research should provide complete guidelines/standard for the wind design of vegetative systems.

The topics of irrigation and maintenance are already quite well developed by the existing standards. A few new recommendations/obligations could be introduced in the standards, as described in the Final Considerations.

As mentioned, one scenario that has very few existing requirements are vegetative systems that are applied to tall buildings; just minimal requirements relating to wind and fire design can be found in the analyzed standards. Considering the specificities of safety issues connected with tall buildings, certain requirements may need specific restrictions and areas that need redefining, depending on the specific standards and research related to existing systems that have already been applied to tall buildings.

For vegetative façades/walls, a huge effort must be done, considering that no extensive testing has been performed and there is a wide range of systems employed to achieve a vertical green system with vastly different characteristics (i.e., modular wall panels, a hanging garden, hydroponics, etc.). Due to the increasing popularity of green façades/walls, it is important that an effort is made to update the standards.

The improvement of existing research, testing, and requirements by international, interdisciplinary experts is necessary for creating a globally accepted standards, to increase the amount of vegetative systems used, while also encouraging the continued advancement in new systems for greening buildings.

Bibliography

Building Codes/Safety Codes

International Code Council, 2012. IEBC-International Existing Building Code.

International Code Council, 2018. IBC-International Building Code.

International Code Council, 2015. IGCC-International Green Construction Code.

NFPA-National Fire Protection Association, 2012. 285: Standard Fire Test Method for Evaluation of Fire Propagation Characteristics of Exterior Non-Load-Bearing Wall Assemblies Containing Combustible Components.

NFPA-National Fire Protection Association, 2015. 5000: Building Construction and Safety Code.

NFPA-National Fire Protection Association, 2015. 101: Life Safety Code.

City Codes/Regional Guidelines

Groundwork South Yorkshire, 2014. The GRO Green Roofs Code.

State of Victoria through the Department of Environment and Primary Industry, 2014. Growing Green Guide. A Guide to Green Roofs, Walls and Facades in Melbourne and Victoria, Australia.

Toronto Green Roof Construction Standard-TGRCS, Supplementary Guidelines.

Toronto Municipal Code, 2013. Chapter 492: Green roofs.

Guidelines/Manuals

CIBSE-Chartered Institution of Building Services Engineers, 2013. Guidelines for the Design and Application of Green Roof System.

CIBSE-Chartered Institution of Building Services Engineers, 2007. Knowledge Series 11 Green Roofs.

FLL-Forschungsgesellschaft Landschaftsentwicklung Landschaftsbau e.V. – The Landscaping and Landscape Development Research Society e.V., 2018. Guidelines for the Planning, Construction and Maintenance of Green Roofing.

GRHC-Green Roofs for Healthy Cities, 2011. Advanced Green Roof Maintenance.

GRHC-Green Roofs for Healthy Cities, 2014. Green Walls 101: Systems Overview and Design.

Ispra-Istituto Superiore per la Protezione e la Ricerca Ambientale, 2012. Verde Pensile: prestazioni di sistema e valore ecologico. Manuali e Linee Guida 78/2012.

Insurance Data Sheets

FM Global, 2011. Property Loss Prevention Data Sheets 1-35 – Green Roof Systems.

Research Reports

Department for Communities and Local Government-London, 2013. Fire Performance of Green Roofs and Walls.

FPRF-Fire Protection Research Foundation, 2012. Fire Safety Challenges of Green Buildings: Final Report.

National Association of State Fire Marshals: Tidwell, Jim and Jack J.Murphy, 2010. Bridging the Gap: Fire Safety and Green Buildings. A Fire and Building Safety Guide to Green Construction, 2010.

Standards

ANSI/SPRI-American National Standard Institution/Single Ply Roofing Industry, 2017. ANSI/SPRI VF-1 External Fire Design Standard for Vegetative Roofs.

ANSI/SPRI-American National Standard Institution/Single Ply Roofing Industry, 2016. ANSI/SPRI RP-14, Wind Design Standard for Vegetative Roofing System.

ASTM International-American Society for Testing and Materials International, 2016. E108-11, Standard Test Methods for Fire Tests of Roof Coverings.

ASTM International-American Society for Testing and Materials International, 2015. ASTM E2397 Standard Practice for Determination of Dead Loads and Live Loads Associated with Vegetative (Green) Roof Systems.

ASTM International-American Society for Testing and Materials International, 2015. ASTM E2398 Standard Test Method for Water Capture and Media Retention of Geocomposite Drain Layers for Vegetative (Green) Roof Systems.

ASTM International-American Society for Testing and Materials International, 2015. ASTM E2400 Standard Guide for Selection, Installation, and Maintenance of Plants for Green Roof Systems.

ASTM International-American Society for Testing and Materials International, 2014. ASTM E2777 Standard Guide for Vegetative (Green) Roof Systems.

Austrian Standards, 2010. Önorm L 1131:2010 06 01 Gartengestaltung und Landschaftsbau-Begrünung von Dächern und Decken auf Bauwerken-Anforderungen an Planung, Ausführung und Erhaltung (Horticulture and Landscaping-Green Area on Roofs and Ceilings of Buildings – Directives for Planning, Building and Maintenance).

CSA Group-Canadian Standards Association, 2015. A123.24-15 Standard Test Method for Wind Resistance of Modular Vegetative Roof Assembly.

European Committee for Standardization, 2007/2009. EN13501-1:2007+A1:2009 European Fire Classification of Construction Products and Building Elements.

MCCAA-Malta Competition and Consumers Affairs Authority, Standards and Metrology Institute, SM 3700, 2017. Green Roofs – Criteria for the planning, construction, control and maintenance of Green Roofs.

SIA-Schweizerische Ingenieur und Architektenverein, 2013. SIA 312:2013 Begrünung von Dächern (Greening Roofs).

UNI-Ente Italiano di Normazione, 2015. UNI 11235:2015 Istruzioni per la progettazione, l'esecuzione, il controllo e la manutenzione di coperture a verde (Criteria for Design, Execution, Testing and Maintenance of Roof Gardens).

Index

Note: 'Page numbers followed by "*f*" indicate figures and "*t*" indicate tables.'

Kolibri 360 GmbH
Ferdinand-Jühlke-Straße 7
99095 Erfurt, Deutschland
produktsicherheit@kolibri360.de